Lecture Notes in Physics

T0238076

Springer

Berlin
Heidelberg
New York
Barcelona
Hong Kong
London
Milan
Paris
Singapore
Tokyo

Physics and Astronomy

ONLINE LIBRARY

http://www.springer.de/phys/

Editorial Policy

The series *Lecture Notes in Physics* (LNP), founded in 1969, reports new developments in physics research and teaching -- quickly, informally but with a high quality. Manuscripts to be considered for publication are topical volumes consisting of a limited number of contributions, carefully edited and closely related to each other. Each contribution should contain at least partly original and previously unpublished material, be written in a clear, pedagogical style and aimed at a broader readership, especially graduate students and nonspecialist researchers wishing to familiarize themselves with the topic concerned. For this reason, traditional proceedings cannot be considered for this series though volumes to appear in this series are often based on material presented at conferences, workshops and schools (in exceptional cases the original papers and/or those not included in the printed book may be added on an accompanying CD ROM, together with the abstracts of posters and other material suitable for publication, e.g. large tables, colour pictures, program codes, etc.).

Acceptance

A project can only be accepted tentatively for publication, by both the editorial board and the publisher, following thorough examination of the material submitted. The book proposal sent to the publisher should consist at least of a preliminary table of contents outlining the structure of the book together with abstracts of all contributions to be included.
Final acceptance is issued by the series editor in charge, in consultation with the publisher, only after receiving the complete manuscript. Final acceptance, possibly requiring minor corrections, usually follows the tentative acceptance unless the final manuscript differs significantly from expectations (project outline). In particular, the series editors are entitled to reject individual contributions if they do not meet the high quality standards of this series. The final manuscript must be camera-ready, and should include both an informative introduction and a sufficiently detailed subject index.

Contractual Aspects

Publication in LNP is free of charge. There is no formal contract, no royalties are paid, and no bulk orders are required, although special discounts are offered in this case. The volume editors receive jointly 30 free copies for their personal use and are entitled, as are the contributing authors, to purchase Springer books at a reduced rate. The publisher secures the copyright for each volume. As a rule, no reprints of individual contributions can be supplied.

Manuscript Submission

The manuscript in its final and approved version must be submitted in camera-ready form. The corresponding electronic source files are also required for the production process, in particular the online version. Technical assistance in compiling the final manuscript can be provided by the publisher's production editor(s), especially with regard to the publisher's own Latex macro package which has been specially designed for this series.

Online Version/ LNP Homepage

LNP homepage (list of available titles, aims and scope, editorial contacts etc.):
http://www.springer.de/phys/books/lnpp/
LNP online (abstracts, full-texts, subscriptions etc.):
http://link.springer.de/series/lnpp/

Rolf Haug   Herbert Schoeller   (Eds.)

Interacting Electrons in Nanostructures

 Springer

Editors

Prof. Rolf Haug
Abteilung Nanostrukturen
Institut für Festkörperphysik
Universität Hannover
Appelstr. 2
30167 Hannover, Germany

Prof. Herbert Schoeller
Institut für Theoretische Physik
Lehrstuhl A
RTWH Aachen
52056 Aachen, Germany

Cover picture: see contribution by König et al. in this volume

---

Library of Congress Cataloging-in-Publication Data applied for.

Die Deutsche Bibliothek - CIP-Einheitsaufnahme

Interacting electrons in nanostructures / Rolf Haug ; Herbert Schoeller
(ed.). - Berlin ; Heidelberg ; New York ; Barcelona ; Hong Kong ; London ;
Milan ; Paris ; Singapore ; Tokyo : Springer, 2001
  (Lecture notes in physics ; Vol. 579)
  (Physics and astronomy online library)

ISSN 0075-8450

ISBN 978-3-642-07587-2   e-ISBN 978-3-540-45532-5

a member of BertelsmannSpringer Science+Business Media GmbH http://www.springer.de
© Springer-Verlag Berlin Heidelberg 2010

Camera-data conversion by Steingraeber Satztechnik GmbH Heidelberg
Cover design: *design & production*, Heidelberg

Printed on acid-free paper

Preface

Nanostructures are among the most actively studied systems in present day research. Theoretical and experimental studies of these small systems have uncovered remarkable phenomena due to fundamental interactions and the reduction in dimensionality. Such nanostructures can be made out of well known materials such as metals or semiconductors by lithographic procedures. In addition, novel material systems, e.g. semiconducting heterostructures, or even molecular-like materials, e.g. carbon nanotubes, can represent nanostructures in which electron–electron interactions are of importance.

The discoveries of quantum Hall effects offered an initial glimpse into the striking phenomena that may occur in these low-dimensional structures. Extensive research since then, made possible by enormous advances in fabrication technologies, have revealed a large class of unexpected behaviors in nanostructures. The rapid technological progress in the last two decades has led to great excitement among both applied and basic physicists. The hope that miniaturization of such devices will lead to a new age of smaller and faster computers has motivated extensive work, trying to understand the basic physics involved, an imperative step towards any future application.

During this research many surprising experimental effects have been found, including the integer and the fractional quantum Hall effect, weak localization, conductance quantization in point contacts, universal conductance fluctuations and Coulomb oscillations in quantum dots. In many cases the experimental observations were unexpected. Theoretical progress was important in facilitating understanding of the effects discovered. A deep understanding of the observed phenomena and a consistent picture of the underlying physics are not only important from the basic research point of view, but will also prove crucial for our ability to design and control the operation of novel devices. All the discovered, intriguing effects, which are understood as arising from microscopic quantum interactions, represent remarkable macroscopic scale quantum phenomena. Current research offers some answers to fundamental issues, and it reveals many puzzles that remain to be explained. Studies of these phenomena are among the most active areas of research at the advancing frontiers of physics. In addition to the fundamental issues involved, research is also driven by commercial aspects since it is assumed that present-day silicon technology will reach some ultimate limits within the next 20 years. Therefore, worldwide efforts are underway in search of alternatives for improvement or replacement of silicon technology.

For example, the Coulomb oscillations in quantum dots have demonstrated the ability to change the current through such a device by four or five orders of magnitude by an externally controllable parameter, e.g. a gate voltage. At almost any gate voltage an electron must have a finite energy in order to overcome the on-dot Coulomb repulsion and to be able to tunnel into the dot. Therefore, the conductance through the device is suppressed. The exceptions are the points of charge degeneracy. At these points, two charge states of the dot have the same energy, and an electron can hop on and off the dot without paying an energy penalty. This dramatic effect promptly led to suggestions that the device could be used as a single-electron tunneling transistor. Such a system with a quantum dot coupled via tunnel junctions to two leads serves simultaneously as a model system for electronic transport through strongly interacting mesoscopic systems. Quantum dots are zero-dimensional systems with a fully discrete level spectrum justifying calling them 'artificial atoms'. By changing a gate voltage the number of electrons confined in a quantum dot can be varied in a defined way. Thus, a single-electron transistor with a quantum dot provides a well-controlled object for studying quantum many-body physics. In many respects, such a device resembles an atom embedded into a Fermi sea of electrons.

By increasing the system size one can move from zero-dimensional quantum dots to one-dimensional quantum wires. The electronic properties of one-dimensional systems have attracted considerable attention in theory and experiment. Starting with the work of Tomonaga in 1950 it has become clear that the electron–electron interaction destroys the sharp Fermi surface and leads to a breakdown of the ubiquitous Fermi liquid theory pioneered by Landau for systems of higher dimensionality. Recent years have seen a large number of theoretical and experimental works to study the consequences of these interactions in quantum wires, where ballistic conduction has also been observed. Future applications of these quantum wires might be connected with the need for perfect leads in new electronic devices.

Another alternative to standard silicon technology could conceivably be based on molecular electronics. This field understood as information processing via single molecules, was initiated in the 1970s. As time passed, the manipulation of single molecules appeared to be more difficult than expected and interest faded away. Nowadays, molecular electronics is experiencing a revival, since technology has progressed with the invention of scanning techniques, etc. Connecting a single molecule was proven to be feasible. A second reason for this renewed interest and exciting progress is the discovery of very stable, long and at the same time electronically very versatile molecules: carbon nanotubes. Within a few years, many applications of carbon nanotubes were proposed. They cover quite diverse fields ranging from novel material for storage of hydrogen, application as artificial muscle through to molecular electronics. On the other hand, from a more fundamental point of view it became clear that carbon nanotubes can serve as perfect model systems for the study of interacting electrons in nanostructures. Carbon nanotubes appear to be perfect one-dimensional systems which can be designed to be a one-dimensional metal or a semiconductor.

Applications of quantum mechanics in nanostructures are always related to the manipulation of wave functions. People envisage using quantum mechanics in totally new computer concepts by calculating with entangled states. Several proposals were made to also use nanostructures in such quantum computing concepts. The possibility of manipulating wave functions relies on a certain coherence of the wave function. Coherence is especially important in such quantum computing concepts, but it is at the same time a very fundamental problem. Therefore, the coherence of wave functions, possible dephasing and relaxation processes have been studied theoretically and experimentally in nanostructures in recent years.

Total wave functions carry orbital and spin components. By involving the effects of interacting spins, additional degrees of freedom emerge for the design of new devices. One example of an effect caused by interacting spin systems is the recently discovered Kondo effect in quantum dots. It has been known for many years that the interaction between the spin of a magnetic impurity and the spins of the electrons in a metal can lead to anomalies in transport. Similar effects were predicted for the interaction between the spin of an electron in a quantum dot and the electrons in the leads. Recently, several groups have succeeded in observing this Kondo effect in tunneling through a quantum dot.

Spin effects are also important in ferromagnets. In current information technology, semiconductors and ferromagnets play complementary roles. The low carrier densities in semiconductors facilitate modulation of electronic transport properties by doping or external gates, whereas ferromagnets allow large changes in magnetic and transport properties through the reorientation of magnetic moments in small magnetic fields. Semiconductors are traditionally used for information processing, whereas ferromagnets have their application in storage devices. Recently, the prospect of synergies between electronic and magnetic manipulation of transport properties has raised interest in spin electronics in semiconductors. This interest is also driven by great improvements in material science with the success in growing ferromagnetic semiconductors. New concepts for devices based on spin effects are appearing and a new term has even been coined: spintronics.

Proposals for new device concepts also make use of the interplay between the electronic and the mechanical properties of a system. Such ideas are possible due to the tremendous advances in nanotechnology which allow one to purpose design mechanical properties of small systems. Again it will be important to treat these small systems in a fully quantum mechanical way. Quantum mechanics brings a new quality to the system, so that these nano-electromechanical systems are not just extensions of micro-mechanics. Nano-electromechanical systems are investigated because of their promising features with regard to applications in sensors and communication technology.

This book is based on contributions of a workshop held in Bad Honnef, Germany, from 13–16 June 2000. At the workshop key scientists convened to present their most recent research and thus created a forum for the in-depth discussion of exciting new results. Several of the talks resulted in not only lively but truly controversial discussions. For this book a selection of 12 topics from the work-shop was chosen. Leading scientists have contributed carefully prepared chapters to cover the most interesting aspects of interaction effects in nanostructures. In a few cases it was even possible to bring together experimentalists and theoreti-cians enabling an especially comprehensive view of a particular topic. The 12 articles are arranged into the six parts of this book. The first part concentrates on electronic transport through quantum dots. It is followed by a part about interaction effects in quantum wires. Electronic transport through molecules is the subject of the next part. Effects of phase coherence are discussed in the following part. A further part covers spin electronics and magnetism, and the last demonstrates the connection to nano-mechanics. In this way the current, most exciting problems in the field of nanostructures are well represented in this book.

Hannover and Karlsruhe,   *Rolf Haug*
March 2001   *Herbert Schoeller*

Contents

Part V   Spintronics and Magnetism

Part VI   Nanomechanics

Part I

Quantum Dots

Magnetic-Field-Induced Kondo Effects in Coulomb Blockade Systems

M. Pustilnik[1], L. I. Glazman[1], D. H. Cobden[2], and L. P. Kouwenhoven[3]

[1] Theoretical Physics Institute, University of Minnesota,
   116 Church St. SE, Minneapolis, MN 55455, USA
[2] Department of Physics, University of Warwick, Coventry, CV4 7AL, UK
[3] Department of Applied Physics and ERATO Mesoscopic Correlation Project,
   Delft University of Technology, P. O. Box 5046, 2600 GA Delft, the Netherlands

Abstract. We review the peculiarities of transport through a quantum dot caused by the spin transition in its ground state. Such transitions can be induced by a magnetic field. Tunneling of electrons between the dot and leads mixes the states belonging to the ground state manifold of the dot. Unlike the conventional Kondo effect, this mixing, which occurs only at the singlet-triplet transition point, involves both the orbital and spin degrees of freedom of the electrons. We present theoretical and experimental results that demonstrate the enhancement of the conductance through the dot at the transition point.

1 Introduction

Quantum dot devices provide a well–controlled object for studying quantum many-body physics. In many respects, such a device resembles an atom imbedded into a Fermi sea of itinerant electrons. These electrons are provided by the leads attached to the dot. The orbital mixing in the case of quantum dot corresponds to the electron tunneling through the junctions connecting the dot with leads. Voltage V_g applied to a gate – an electrode coupled to the dot capacitively – allows one to control the number of electons N on the dot. Almost at any gate voltage an electron must have a finite energy in order to overcome the on-dot Coulomb repulsion and tunnel into the dot. Therefore, the conductance of the device is suppressed at low temperatures (Coulomb blockade phenomenon[1]). The exceptions are the points of charge degeneracy. At these points, two charge states of the dot have the same energy, and an electron can hop on and off the dot without paying an energy penalty. This results in a periodic peak structure in the dependence of the conductance G on V_g. Away from the peaks, in the Coulomb blockade valleys, the charge fluctuations are negligible, and the number of electrons N is integer.

Every time N is tuned to an odd integer, the dot must carry a half-integer spin. In the simplest case, the spin is $S = 1/2$, and is due to a single electron residing on the last occupied discrete level of the dot. Thus, the quantum dot behaves as $S = 1/2$ magnetic impurity imbedded into a tunneling barrier between two massive conductors. It is known[2] since mid-60's that the presence of such impurities leads to zero-bias anomalies in tunneling conductance[3], which are adequately explained[4] in the context of the Kondo effect[5]. The advantage of the new experiments[6] is in full control over the "magnetic impurity" responsible for the effect. For example, by varying the gate voltage, N can be changed. Kondo effect results in the increased low–temperature conductance only in the odd–N valleys. The even–N valleys nominally correspond to the $S = 0$ spin state (non-magnetic impurity), and the conductance decreases with lowering the temperature.

Unlike the real atoms, the energy separation between the discrete states in a quantum dot is fairly small. Therefore, the $S = 0$ state of a dot with even number of electrons is much less robust than the corresponding ground state of a real atom. Application of a magnetic field in a few–Tesla range may result in a transition to a higher-spin state. In such a transition, one of the electrons residing on the last doubly–occupied level is promoted to the next (empty) orbital state. The increase in the orbital energy accompanying the transition is compensated by the decrease of Zeeman and exchange energies. At the transition point, the ground state of the dot is degenerate. Electron tunneling between the dot and leads results in mixing of the components of the ground state. Remarkably, the mixing involves spin as well as orbital degrees of freedom. In this paper we demonstrate that the mixing yields an enhancement of the low–temperature conductance through the dot. This enhancement can be viewed as the magnetic–field–induced Kondo effect.

We present the model and theory of electron transport in the conditions of the field-induced Kondo effect in Section 2. The experimental manifestations of the transition observed on GaAs vertical quantum dots and carbon nanotubes are described in Section 3.

2   The Model

We will be considering a confined electron system which does not have special symmetries, and therefore the single-particle levels in it are non-degenerate. In addition, we assume the electron-electron interaction to be relatively weak (the gas parameter $r_s \lesssim 1$). Therefore, discussing the ground state, we concentrate on the transitions which involve only the lowest-spin states. In the case of even number of electrons, these are states with $S = 0$ or $S = 1$. At a sufficiently large level spacing $\delta \equiv \epsilon_{+1} - \epsilon_{-1}$ between the last occupied (-1) and the first empty orbital level $(+1)$, the ground state is a singlet at $B = 0$. Finite magnetic field affects the orbital energies; if it reduces the difference between the energies of the said orbital levels, a transition to a state with $S = 1$ may occur, see Fig. 1. Such a transition involves rearrangement of two electrons between the levels $n = \pm 1$. Out of the six states involved, three belong to a triplet $S = 1$, and three others are singlets $(S = 0)$. The degeneracy of the triplet states is removed only by Zeeman energy. The singlet states, in general, are not degenerate with each other. To describe the transition between a singlet and the triplet in the ground state, it is sufficient to consider the following Hamiltonian:

$$H_{\mathrm{dot}} = \sum_{ns} \epsilon_n d^\dagger_{ns} d_{ns} - E_S S^2 - E_Z S^z + E_C \left(N - \mathcal{N}\right)^2. \tag{1}$$

Here, $N = \sum_{s,n} d^\dagger_{ns} d_{ns}$ is the total number of electrons occupying the levels $n = \pm 1$, operator $S = \sum_{nss'} d^\dagger_{ns} \left(\sigma_{ss'}/2\right) d_{ns'}$ is the corresponding total spin (σ are the Pauli matrices), and the parameters E_S, $E_Z = g\mu_B B$, and E_C are the exchange, Zeeman, and charging energies respectively [7]. We restrict our attention to the very middle of a Coulomb blockade valley with an even number of electrons in the dot (that is modelled by setting the dimensionless gate voltage $\mathcal{N}$ to $\mathcal{N} = 2$). We assume that the level spacing δ is tunable, $e.g.$, by means of a magnetic field B: $\delta = \delta(B)$, and that $\delta(0) > 2E_S$ (which ensures that the dot is non-magnetic for $B = 0$).

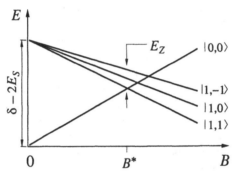

Fig. 1. Typical picture of the singlet-triplet transition in the ground state of a quantum dot.

The lowest–energy singlet state and the three components of the competing triplet state can be labeled as $|S, S^z\rangle$ in terms of the total spin S and its z–projection S_z,

$$|1,1\rangle = d^\dagger_{+1\uparrow} d^\dagger_{-1\uparrow} |0\rangle,$$
$$|1,-1\rangle = d^\dagger_{+1\downarrow} d^\dagger_{-1\downarrow} |0\rangle, \tag{2}$$
$$|1,0\rangle = \frac{1}{\sqrt{2}} \left(d^\dagger_{+1\uparrow} d^\dagger_{-1\downarrow} + d^\dagger_{+1\downarrow} d^\dagger_{-1\uparrow} \right) |0\rangle,$$
$$|0,0\rangle = d^\dagger_{-1\uparrow} d^\dagger_{-1\downarrow} |0\rangle,$$

where $|0\rangle$ is the state with the two levels empty. According to (1), the energies of these states satisfy

$$E_{|S,S^z\rangle} - E_{|0,0\rangle} = K_0 S - E_Z S^z, \tag{3}$$

where $K_0 = \delta - 2E_S$. Since $\delta > 2E_S$, the ground state of the dot at $B = 0$ is a singlet $|0,0\rangle$. Finite field shifts the singlet and triplet states due to the orbital effect, and also leads to Zeeman splitting of the components of the triplet. As B is varied, the level crossings occur (see Fig. 1). The first such crossing takes place at $B = B^*$, satisfying the equation

$$\delta(B^*) - E_Z(B^*) = 2E_S. \tag{4}$$

At this point, the two states, $|0,0\rangle$ and $|1,1\rangle$, form a doubly degenerate ground state, see Fig. 1.

If leads are attached to the dot, the dot-lead tunneling results in the hybridization of the degenerate (singlet and triplet) states. The characteristic energy scale T_0 associated with the hybridization can be in different relations with the Zeeman splitting at field $B = B^*$.

If $E_Z(B^*) \ll T_0$, then the Zeeman splitting between the triplet states can be neglected, and at the $B = B^*$ point all *four* states (2) can be considered as degenerate. Theory for this case is presented below in Section 2.3. This limit adequately describes a quantum dot formed in a two-dimensional electron gas (2DEG) at the GaAs-AlGaAs interface, subject to a magnetic field, see Section 3.1. Energy E_Z can be neglected due to the smallness of the electron g-factor in GaAs.

Alternatively, the orbital effect of the magnetic field (B-dependence of δ) may be very weak due to the reduced dimensionality of the system, while the g-factor is not suppressed, yielding an appreciable Zeeman effect even in a magnetic field of a moderate strength. This limit of the theory, see Section 2.4, corresponds to single-wall carbon nanotubes, which have very small widths of about 1.4 nm and $g = 2.0$. Measurements with carbon nanotubes are presented in Section 3.2.

In order to study the transport problem, we need to introduce into the model the Hamiltonian of the leads and a term that describes the tunneling. We choose them in the following form:

$$H_l = \sum_{anks} \xi_k c^\dagger_{anks} c_{anks}, \tag{5}$$

$$H_T = \sum_{ann'ks} t_{ann'} c^\dagger_{anks} d_{n's} + \text{H.c.} \tag{6}$$

Here $\alpha = R, L$ for the right/left lead, and $n = \pm 1$ for the two orbitals participating in the singlet-triplet transition; k labels states of the continuum spectrum in the leads, and s is the spin index. In writing (5)-(6), we had in mind the vertical dot device, where the potential creating lateral confinement of electrons most probably does not vary much over the thickness of the dot[8]. Therefore we have assumed that the electron orbital motion perpendicular to the axis of the device can be characterised by the same quantum number n inside the dot and in the leads. Presence of two orbital channels $n = \pm 1$ is important for the description of the Kondo effect at the singlet-triplet transition, that is, when the orbital effect of the magnetic field dominates. In the opposite case of large Zeeman splitting, the problem is reduced straightforwadly to the single-channel one, as we will see in Section 2.4 below.

2.1 Effective Hamiltonian

We will demonstrate the derivation of the effective low-energy Hamiltonian under the simplifying assumption [9],[10]

$$t_{\alpha nn'} = t_\alpha \delta_{nn'}. \tag{7}$$

This assumption, on one hand, greatly simplifies the calculations, and, on the other hand, is still general enough to capture the most important physical properties [11].

It is convenient to begin the derivation by performing a rotation[12] in the R-L space

$$\begin{pmatrix} \psi_{nks} \\ \phi_{nks} \end{pmatrix} = \frac{1}{\sqrt{t_L^2 + t_R^2}} \begin{pmatrix} t_R & t_L \\ -t_L & t_R \end{pmatrix} \begin{pmatrix} c_{Rnks} \\ c_{Lnks} \end{pmatrix}, \tag{8}$$

after which the ϕ field decouples:

$$H_T = \sqrt{t_L^2 + t_R^2} \sum_{nks} \psi_{nks}^\dagger d_{ns} + \text{H.c.} \tag{9}$$

The differential conductance at zero bias G can be related, using Eq. (8), to the amplitudes of scattering $A_{ns \to n's'}$ of the ψ–particles

$$G = \lim_{V \to 0} dI/dV = \frac{e^2}{h} \left(\frac{2t_L t_R}{t_L^2 + t_R^2} \right)^2 \sum_{nn'ss'} |A_{ns \to n's'}|^2. \tag{10}$$

The next step is to integrate out the virtual transitions to the states with $\mathcal{N} \pm 1$ electrons by means of the Schrieffer-Wolff transformation or, equivalently, by the Brillouin–Wigner perturbation theory. This procedure results in the effective low-energy Hamiltonian in which the transitions between the states (2) are described by of the operators

$$S_{nn'} = \mathcal{P} \sum_{ss'} d_{ns}^\dagger \frac{\sigma_{ss'}}{2} d_{n's'} \mathcal{P},$$

where $\mathcal{P} = \sum_{S,S^z} |S, S^z\rangle\langle S, S^z|$ is the projection operator onto the system of states (2). The operators $S_{nn'}$ may be conveniently written in terms of two fictitious 1/2-spins $S_{1,2}$. The idea of mapping comes from the one-to-one correspondence between the set of states (2) and the states of a two-spin system:

$$|1,1\rangle \iff |\uparrow_1 \uparrow_2\rangle, \quad |1,-1\rangle \iff |\downarrow_1 \downarrow_2\rangle,$$

$$|1,0\rangle \iff \frac{1}{\sqrt{2}} \left(|\uparrow_1 \downarrow_2\rangle + |\downarrow_1 \uparrow_2\rangle \right),$$

$$|0,0\rangle \iff \frac{1}{\sqrt{2}} \left(|\uparrow_1 \downarrow_2\rangle - |\downarrow_1 \uparrow_2\rangle \right).$$

We found the following relations:

$$S_{nn} = \frac{1}{2}(S_1 + S_2) = \frac{1}{2}S_+,$$

$$\sum_n S_{-n,n} = \frac{1}{\sqrt{2}}(S_1 - S_2) = \frac{1}{\sqrt{2}}S_-, \tag{11}$$

$$\sum_n in S_{-n,n} = \sqrt{2}[S_1 \times S_2] = \sqrt{2}\mathbf{T}.$$

In terms of $S_{1,2}$, the effective Hamiltonian takes the form:

$$H = \sum_{nks} \xi_k \psi^\dagger_{nks}\psi_{nks} + K(S_1 \cdot S_2) - E_Z S^z_+ + \sum_n H_n, \tag{12}$$

$$H_n = J(s_{nn} \cdot S_+) + V n \rho_{nn}(S_1 \cdot S_2) \tag{13}$$

$$+ \frac{I}{\sqrt{2}}[(s_{-n,n} \cdot S_-) + 2in(s_{-n,n} \cdot \mathbf{T})].$$

Here we introduced the particle and spin densities in the continuum:

$$\rho_{nn} = \sum_{kk's} \psi^\dagger_{nks}\psi_{nk's}, \quad s_{nn'} = \sum_{kk'ss'} \psi^\dagger_{nks}\frac{\sigma_{ss'}}{2}\psi_{n'k's'}.$$

The bare values of the coupling constants are

$$J = I = 2V = 2(t_L^2 + t_R^2)/E_C. \tag{14}$$

Note that the Schrieffer-Wolff transformation also produces a small correction to the energy gap Δ between the states $|1,1\rangle$ and $|0,0\rangle$,

$$\Delta = E_{|1,1\rangle} - E_{|0,0\rangle} = K - E_Z, \tag{15}$$

so that K differs from its bare value K_0, see (3). However, this difference is not important, since it only affects the value of the control parameter at which the singlet-triplet transition occurs, but not the nature of the transition.

We did not include into (1)-(13) the free-electron Hamiltonian of the ϕ-particles [see Eq. (8)], as well as some other terms, that are irrelevant for the low energy renormalization. The contribution of these terms to the conductance is featureless at the energy scale of the order of T_0 (see the next section), where the Kondo resonance develops.

At this point, it is necessary to discuss some approximations tacitly made in the derivation of (1)-(13). First of all, we entirely ignored the presence of many energy levels in the dot, and took into account the low-energy multiplet (2) only. The multi-level structure of the dot is important at the energies above δ, while the Kondo effect physics emerges at the energy scale well below the single-particle level spacing [13]. The high-energy states result merely in a renormalization of the parameters of the effective low-energy Hamiltonian. One only needs to consider this renormalization for deriving the relation between the parameters t_L

and t_R of the low-energy Hamiltonian (1), (5) and (6) and the "bare" constants of the model defined in a wide bandwidth ϵ_F. On the other hand, using the effective low-energy Hamiltonian, one can calculate, in principle, the observable quantities such as conductance $G(T)$ and other susceptibilities of the system at low temperatures $(T \ll \delta)$, and establish the relations between them, which is our main goal.

Note that the Hamiltonian (1)-(13) resembles that of the two-impurity Kondo model, for which $H_n = J_n (s_{nn} \cdot S_+) + I (s_{-n,n} \cdot S_-)$ and the parameter K characterizes the strength of the RKKY interaction [14]. It is known that the two-impurity Kondo model may undergo a phase transition at some special value of K [14]. At this point, the system may exhibit non-Fermi liquid properties. However, one can show [10], using general arguments put forward in [14], that the model (1)-(13) does not have the symmetry that warrants the existence of the non Fermi liquid state. This allows one to apply the local Fermi liquid description [15] to study the properties of the system at $T = 0$. In the next section, we will concentrate on the experimentally relevant perturbative regime.

2.2   Scaling Analysis

To calculate the differential conductance in the leading logarithmic approximation, we apply the "poor man's" scaling technique [16]. The procedure consists of a perturbative elimination of the high-energy degrees of freedom and yields the and yields the set of scaling equations

$$
\begin{aligned}
dJ/d\mathcal{L} &= \nu \left(J^2 + I^2 \right), \\
dI/d\mathcal{L} &= 2\nu I \left(J + V \right), \\
dV/d\mathcal{L} &= 2\nu I^2
\end{aligned}
\qquad (16)
$$

for the renormalization of the coupling constants with the decrease of the high energy cutoff D. Here $\mathcal{L} = \ln(\delta/D)$, and ν is the density of states in the leads; the initial value of D is $D = \delta$, see the discussion after Eq. (15). The initial conditions for (16), $J(0)$, $I(0)$, and $V(0)$ are given by Eq. (14). The scaling procedure also generates non-logarithmic corrections to K. In the following we absorb these corrections in the re-defined value of K. Equations (16) are valid in the perturbative regime and as long as

$$
D \gg |K|, E_Z, T.
$$

At certain value of $\mathcal{L}$, $\mathcal{L} = \mathcal{L}_0 = \ln(\delta/T_0)$, the inverse coupling constants simultaneously reach zero:

$$
1/J (\mathcal{L}_0) = 1/I (\mathcal{L}_0) = 1/V (\mathcal{L}_0) = 0.
$$

This defines the characteristic energy scale of the problem:

$$
T_0 = \delta \exp \left[-\tau_0 / \nu J \right].
\qquad (17)
$$

Here τ_0 is a parameter that depends on the initial conditions and should be found numerically. We obtained $\tau_0 = 0.36$ (see Fig. 2).

It is instructive to compare T_0 with the Kondo temperature T_K^{odd} in the adjacent Coulomb blockade valleys with $N = odd$. In this case, only electrons from one of the two orbitals $n = \pm 1$ are involved in the effective Hamiltonian, which takes the form of the 1-channel $S = 1/2$ Kondo model with the exchange amplitude $J_{\mathrm{odd}} = 4(t_L^2 + t_R^2)/E_C = 2J$, see Eq. (14). Therefore, T_K^{odd} is given by the same expression (3.2) as T_0, but with $\tau_0 = 1/2$. For realistic values of the parameters $T_0 = 300 \ mK$, $\delta = 3 \ meV$ we obtain $T_K^{\mathrm{odd}} \approx 120 \ mK$. This estimate is in a reasonable agreement with the experimental data, see Section 3.1 below.

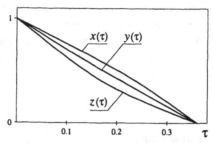

Fig. 2. Numerical solution of the scaling equations. The RG equations (16) are rewritten in terms of the new variable $\tau = \nu J(0) \ln(\delta/D)$ and the new functions $x(\tau) = J(0)/J(\tau)$, $y(\tau) = I(0)/I(\tau)$, $z(\tau) = V(0)/V(\tau)$ as $dx/d\tau = -(1 + x^2/y^2)$, $dy/d\tau = -(2y/x + y/z)$, $dz/d\tau = -4z^2/y^2$. The three functions reach zero simultaneously at $\tau = \tau_0 = 0.36$.

The solution of the RG equations (16) can now be expanded near $\mathcal{L} = \mathcal{L}_0$. To the first order in $\mathcal{L}_0 - \mathcal{L} = \ln D/T_0$, we obtain

$$\frac{1}{\nu J(\mathcal{L})} = \frac{\sqrt{\lambda}}{\nu I(\mathcal{L})} = \frac{\lambda - 1}{2\nu V(\mathcal{L})} = (\lambda + 1)\ln(D/T_0), \tag{18}$$

where

$$\lambda = 2 + \sqrt{5} \approx 4.2.$$

It should be emphasized that, unlike τ_0, the constant λ is universal in the sense that its value is not affected if the restriction (7) is lifted [11].

Eq. (18) can be used to calculate the differential conductance at high temperature $T \gg |K|, E_Z, T_0$. In this regime, the coupling constants are still small, and the conductance is obtained by applying a perturbation theory to the Hamiltonian (1)-(13) with renormalized parameters (18), taken at $D = T$, and using (10). This yields

$$G/G_0 = \frac{A}{[\ln(T/T_0)]^2}, \tag{19}$$

where

$$A = (3\pi^2/8)(\lambda + 1)^{-2}\left[1 + \lambda + (\lambda - 1)^2/8\right] \approx 0.9$$

is a numerical constant, and

$$G_0 = \frac{4e^2}{h} \left(\frac{2t_L t_R}{t_L^2 + t_R^2} \right)^2 .$$ (20)

As temperature is lowered, the scaling trajectory (16) terminates either at $D \sim \max\{|K|, E_Z\} \gg T_0$, or when the system approaches the strong coupling regime $D \sim T_0 \gg |K|, E_Z$. It turns out that the two limits of the theory, $E_Z \ll T_0$ and $E_Z \gg T_0$, describe two distinct physical situations, which we will discuss separately.

2.3 Singlet–Triplet Transition

In this section, we assume that the Zeeman energy is negligibly small compared to all other energy scales. At high temperature $T \gg |K|, T_0$, the conductance is given by Eq. (19). At low temperature $T \lesssim |K|$ and away from the singlet-triplet degeneracy point, $|K| \gg T_0$, the RG flow yielding Eq. (19) terminates at energy $D \sim |K|$. On the *triplet* side of the transition ($K \ll -T_0$), the two spins $S_{1,2}$ are locked into a triplet state. The system is described by the effective 2-channel Kondo model with $S = 1$ impurity, obtained from Eqs. (1)-(13) by projecting out the singlet state and dropping the no longer relevant potential scattering term:

$$H_{\text{triplet}} = \sum_{nks} \xi_k \psi_{nks}^\dagger \psi_{nks} + J \sum_n (s_{nn} \cdot S) ;$$ (21)

here J is given by the solution $J(\mathcal{L})$ of Eq. (16), taken at $\mathcal{L} = \mathcal{L}^* = \ln(\delta/|K|)$, which corresponds to $D = |K|$.

As D is lowered below $|K|$, the renormalization of the exchange amplitude J is governed by the standard RG equation [16]

$$dJ/d\mathcal{L} = \nu J^2,$$ (22)

where $\mathcal{L} = \ln(\delta/D) > \mathcal{L}^*$. Eq. (22) is easily integrated with the result

$$1/\nu J(\mathcal{L}) - 1/\nu J(\mathcal{L}^*) = \mathcal{L} - \mathcal{L}^*.$$

This can be also expressed in terms of the running bandwidth D and the Kondo temperature

$$T_k = |K| \exp[-1/\nu J(\mathcal{L}^*)]$$

as $1/\nu J(\mathcal{L}) = \ln(D/T_k)$.

Obviously, T_k depends on $|K|$. Using asymptotes of $J(\mathcal{L})$, see Eq. (18), we obtain the scaling relation

$$T_k/T_0 = (T_0/|K|)^\lambda .$$ (23)

Eq. (23) is valid not too far from the transition point, where the inequality

$$1 \lesssim |K|/T_0 \ll (\delta/T_0)^\mu, \ \mu \approx 0.24$$ (24)

is satisfied. Here, μ is a numerical constant, which depends on τ_0, and therefore is not universal [see the remark after Eq. (18)]. For larger values of $|K|$ (but still smaller than δ), $T_k \propto 1/|K|$ [11]. Finally, for $|K| = \delta$, T_k is given by Eq. (3.2) with $\tau_0 = 1$. According to (23), T_k decreases very rapidly with $|K|$. For example, for $T_0 = 300\ mK$ and $\delta = 3\ meV$ Eq. (23) describes fall of T_k by an order of magnitude within the limits of its validity (24). For $|K| = \delta$ one obtains $T_k \approx 5\ mK$, which is well beyond the reach of the present day experiments.

For a given $|K|$, $T_0 \lesssim |K| \lesssim \delta$, the differential conductance can be cast into the scaling form,

$$G/G_0 = F\left(T/T_k\right) \tag{25}$$

where $F\left(x\right)$ is a smooth function that interpolates between $F\left(0\right) = 1$ and $F\left(x \gg 1\right) = \left(\pi^2/2\right)\left(\ln x\right)^{-2}$. It coincides with the scaled resistivity

$$F(T/T_K) = \rho(T/T_K)/\rho(0)$$

for the symmetric two–channel $S = 1$ Kondo model. The conductance at $T = 0$ (the unitary limit value), G_0, is given above in Eq. (13).

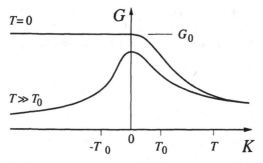

Fig. 3. Linear conductance near a singlet-triplet transition. At high temperature G exhibits a peak near the transition point. At low temperature G reaches the unitary limit at the triplet side of the transition, and decreases monotonously at the singlet side. The two asymptotes merge at $K \gg T, T_0$.

On the *singlet* side of the transition, $K \gg T_0$, the scaling terminates at $D \sim K$, and the low-energy effective Hamiltonian is

$$H_{\text{singlet}} = \sum_{nks} \xi_k \psi^\dagger_{nks} \psi_{nks} - \frac{3}{4} V \sum_n n\rho_{nn},$$

where V is $V(\mathcal{L})$ [see Eq. (16)] taken at $\mathcal{L} = \mathcal{L}^*$. The temperature dependence of the conductance saturates at $T \ll K$, reaching the value

$$G/G_0 = \frac{B}{[\ln(K/T_0)]^2}, \quad B = \left(\frac{3\pi}{8}\frac{\lambda - 1}{\lambda + 1}\right)^2 \approx 0.5. \tag{26}$$

Note that at $T = 0$ Eqs. (25) and (26) predict different dependence on the parameter K which is used for tuning thorugh the transition. At positive K, conductance decreases with the increase of K; at $K \ll -T_0$ conductance $G = G_0$ and does not depend of K. Although there is no reason for the function $G(K)$ to be discontinious [10], it is obviously a *non-analytical* function of K, see Fig. 3.

The above results are for the linear conductance G. At $T = 0$, G is a monotonous function of K, at high temperature $T \gg T_0$ the conductance develops a peak at the singlet-triplet transition point $K = 0$. We now discuss shortly out-of-equilibrium properties. When the system is tuned to the transition point $K = 0$, the differential conductance dI/dV exhibits a peak at zero bias, whose width is of the order of T_0. For finite K the peak splits in two, located at *finite* bias $eV = \pm K$. The mechanism of this effect is completely analogous to the Zeeman splitting of the usual Kondo resonance [4],[2]: $|K|$ is the energy cost of the processes involving a singlet-triplet transition. This cost can be covered by applying a finite voltage $eV = \pm K$, so that the tunneling electron has just the right amount of extra energy to activate the singlet-triplet transition prosesses described by the last two terms in (13). The split peaks gradually disappear at large $|K|$ due to the nonequilibrium-induced decoherence [17],[18].

2.4 Transition Driven by Zeeman Splitting

If the Zeeman energy is large, the RG flow (16) terminates at $D \sim E_Z$. The effective Hamiltonian, valid at the energies $D \lesssim E_Z$ is obtained by projecting (1)-(13) onto the states $|1,1\rangle$ and $|0,0\rangle$. These states differ by a flip of a spin of a single electron (see Fig. 4), and are the counterparts of the spin-up and spin-down states of $S = 1/2$ impurity in the conventional Kondo problem. It is therefore convenient to switch to the notations

$$|1,1\rangle = |\uparrow\rangle, \quad |0,0\rangle = |\downarrow\rangle, \tag{27}$$

and to describe the transitions between the two states in terms of the spin-like operator

$$\widetilde{S} = \frac{1}{2} \sum_{ss'} |s\rangle \sigma_{ss'} \langle s'|,$$

built from the states (27).

Projecting onto the sates (27), we obtain from (1)-(13)

$$\begin{aligned}
H = \ &\sum_{nks} \xi_k \psi^\dagger_{nks} \psi_{nks} + \Delta \widetilde{S}^z \\
&+ \sum_n \left[J s^z_{nn} \left(\widetilde{S}^z + 1/2 \right) + V n \rho_{nn} \left(\widetilde{S}^z - 1/4 \right) \right] \\
&- I \left(s^+_{1,-1} \widetilde{S}^- + \text{H.c.} \right),
\end{aligned} \tag{28}$$

where Δ was introduced above in Eq. (15). It is now convenient to transform (28) to a form which is diagonal in the orbital indexes n. This is achieved simply by relabeling the fields according to

$$\psi_{+1,k,\uparrow} = a_{k,\uparrow}, \quad \psi_{-1,k,\downarrow} = -a_{k,\downarrow}, \tag{29}$$
$$\psi_{-1,k,\uparrow} = b_{k,\uparrow}, \quad \psi_{+1,k,\downarrow} = -b_{k,\downarrow},$$

which yields

$$
\begin{aligned}
H = \ & H_0 + \Delta \widetilde{S}^z \\
& + V_a s_a^z + J_z s_a^z \widetilde{S}^z + \frac{1}{2} J_\perp \left(s_a^+ \widetilde{S}^- + s_a^- \widetilde{S}^+ \right) \\
& + V_b s_b^z + J_z' s_b^z \widetilde{S}^z ,
\end{aligned}
\tag{30}
$$

where H_0 is a free-particle Hamiltonian for a, b electrons, and s_a is the spin density for a electrons, $s_a = \sum_{kk'ss'} a_{ks}^\dagger \left(\sigma_{ss'}/2 \right) a_{k's'}$ (with a similar definition for s_b). The coupling constants in (30),

$$
\begin{aligned}
V_a = (J - V)/2, \quad & J_z = J + 2V, \quad J_\perp = 2I, \\
V_b = (J + V)/2, \quad & J_z' = J - 2V
\end{aligned}
\tag{31}
$$

are expressed through the solutions of the RG equations (16) taken at $\mathcal{L} = \mathcal{L}^{**} = \ln(\delta/E_Z)$.

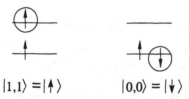

$$|1,1\rangle = |\uparrow\rangle \qquad |0,0\rangle = |\downarrow\rangle$$

Fig. 4. The ground state doublet in case of a large Zeeman splitting. The states $|1,1\rangle$ and $|0,0\rangle$ differ by flipping a spin of a single electron (marked by circles).

The operators in b-dependent part of (30) are not relevant for the low energy renormalization. At low enough temperature (satisfying the condition $\ln(T/T_Z) \ll (\nu J_z')^{-1}, (\nu V_b)^{-1}$, where T_Z is the Kondo temperature), their contribution to the conductance becomes negligible compared to the contribution from the a-dependent terms. This allows us to drop the b-dependent part of (30). Suppressing the (now redundant) subscript of the operators s_a^i, we are left with the Hamiltonian of a one-channel $S = 1/2$ anisotropic Kondo model,

$$
H = H_0 + \Delta \widetilde{S}^z + V_a s^z + J_z s^z \widetilde{S}^z + \frac{J_\perp}{2}(s^+ \widetilde{S}^- + \text{H.c.}).
\tag{32}
$$

Eq. (32) emerged as a limiting case of a more general two-channel model (1)-(13). It should be noticed, however, that the same effective Hamiltonian (32) appears when one starts with the single-channel model from the very beginning [19].

A finite magnetic field singles out the z-direction, so that the spin–rotational symmetry is absent in (1)-(13). This property is preserved in (32). Indeed, even for $J_z = J_\perp$, Eq. (32) contains term $V_\psi s^z$ which has the meaning of a magnetic field acting *locally* on the conduction electrons at the impurity site. The main effect of this term is to produce a correction to Δ, through creating a non-zero expectation value $\langle s^z \rangle$ [20]. This results in a correction to Δ. Fortunately, this correction is not important, since it merely shifts the degeneracy point. In addition, this term leads to insignificant corrections to the density of states [19].

Let us now examine the relation between J_z and $J_\perp$. It follows from Eqs. (14) and (31), that at $E_Z = \delta$ the exchange is isotropic: $J_z = J_\perp$. Moreover, it turns out that if E_Z is so close to δ, that Eqs. (16) can be linearized near the weak coupling fixed point $\mathcal{L} = 0$, the corrections to $J_z, J_\perp$ are such that the isotropy of exchange is preserved:

$$J_z = J_\perp = 2J(0) \left[1 + \frac{3}{4}\nu J(0) \ln(\delta/E_Z)\right], \qquad (33)$$

where $J(0)$ is given by (14). This expression is valid as long as the logarithmic term in the r.h.s. is small: $\nu J(0) \ln(\delta/E_Z) \ll 1$. Using (33) and (3.2), one obtains the Kondo temperature T_Z, which for $J_z = J_\perp$ is given by

$$T_Z = E_Z \exp[-1/\nu J_z] = E_Z (\delta/E_Z)^{3/8} (T_0/\delta)^{1/2\tau_0}$$

Note that for $E_Z = \delta$, T_Z coincides with the Kondo temperature T_K^{odd} in the adjacent Coulomb blockade valleys with odd number of electrons [19], see the discussion after Eq. (3.2) above.

Note that the anisotropy of the exchange merely affects the value of T_Z (which can be written explicitly for arbitrary J_z and $J_\perp$ [21]). In the universal regime (when T approaches T_Z), the exchange can be considered isotropic. This is evident from the scaling equations [16]

$$dJ_z/d\mathcal{L} = \nu J_\perp^2, \; dJ_\perp/d\mathcal{L} = \nu J_z J_\perp, \; \mathcal{L} > \mathcal{L}^{**} \qquad (34)$$

where $\mathcal{L} > \mathcal{L}^{**} = \ln(\delta/E_Z)$, whose solution approaches the line $J_z = J_\perp$ at large $\mathcal{L}$.

According to the discussion above, the term $V_\psi s^z$ in Eq. (32) can be neglected. As a results, (32) acquires the form of the anisotropic Kondo model, with Δ playing the part of the Zeeman splitting of the impurity levels. This allows us to write down the expression for the linear conductance at once. Regardless the initial anisotropy of the exchange constants in Eq. (32), the conductance for $\Delta = 0$ in the universal regime (when T approaches T_Z or lower) is given by

$$G = G_{0Z} f(T/T_Z), \qquad (35)$$

where $f(x)$ is a smooth function interpolating between $f(0) = 1$ and $f(x \gg 1) = (3\pi^2/16)(\ln x)^{-2}$. Function $f(T/T_Z)$ coincides with the scaled resistivity for the one-channel $S = 1/2$ Kondo model and its detailed shape is known from the numerical RG calculations [22]. The conductance at $T = 0$,

$$G_{0Z} = \frac{2e^2}{h} \left(\frac{2t_L t_R}{t_L^2 + t_R^2}\right)^2, \qquad (36)$$

is by a factor of 2 smaller than G_0 [see Eq. (13)]; G_0 includes contributions from two channels and therefore is twice as large as the single-channel result (36). At finite $\Delta \gg T_Z$, the scaling trajectory (34) terminates at $D \sim \Delta$. As a result, at $T \lesssim \Delta$ the conductance is temperature-independent, and for $J_z = J_\perp$

$$G = G_{0Z} f(\Delta/T_Z) = G_{0Z} \frac{3\pi^2/16}{[\ln(\Delta/T_Z)]^2}.$$

The effect of the de-tuning of the magnetic field from the degeneracy point $\Delta \neq 0$ on the differential conductance away from equilibrium is similar to the effect the magnetic field has on the usual Kondo resonance [4],[2]. For example, consider the case, relevant for the experiments on the carbon nanotubes, see section 3.2, when the exchange energy E_S [see Eq. (1)] is negligibly small. When sweeping magnetic field from $B = -\infty$ to $B = +\infty$, the degeneracy between the singlet state of the dot and a component of the triplet is reached twice, at $B = B^*$ and $B = -B^*$, when $|E_Z| \approx \delta$. If the field is tuned to $B = \pm B^*$, then the differential conductance dI/dV has a peak at zero bias. At a finite difference $|B| - |B^*|$ this peak splits in two located at $eV = \pm g\mu_B(|B| - |B^*|)$.

3   Experiments

3.1   GaAs Quantum Dots

Here, we discuss the case of a quantum dot with $N = even$ in a situation where the last two electrons occupy either a spin singlet or a spin triplet state. The transition between singlet and triplet state is controlled with an external magnetic field. The range of the magnetic field is small ($B \sim 0.2\ T$, $g\mu_B B \sim 5\ \mu V$) such that the Zeeman energy can be neglected and that the triplet state is fully degenerate[23]. The theory for this situation was described in section 2.3.

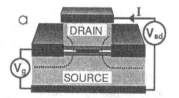

Fig. 5. (a) Cross-section of rectangular quantum dot. The semiconductor material consists of an undoped AlGaAs(7nm)/InGaAs(12nm)/AlGaAs(7nm) double barrier structure sandwiched between n-doped GaAs source and drain electrodes. A gate electrode surrounds the pillar and is used to control the electrostatic confinement in the quantum dot. A dc bias voltage, V, is applied between source and drain and current, I, flows vertically through the pillar. The gate voltage, V_g, can change the number of confined electrons, N, one-by-one. A magnetic field, B, is applied along the vertical axis. (b) Scanning electron micrograph of a quantum dot with dimensions $0.45 \times 0.6\ \mu m^2$ and height of $\sim 0.5\ \mu m$.

The quantum dot has the external shape of a rectangular pillar (see Fig. 5) and an internal confinement potential close to a two-dimensional ellipse [8]. The tunnel barriers between the quantum dot and the source and drain electrodes are thinner than in other devices such that higher-order tunneling processes are enhanced. Fig. 6 shows the linear response conductance G versus gate voltage V_g, and magnetic field B. Dark regions have low conductance and correspond to the regimes of Coulomb blockade for $N = 3$ to 10. Light stripes represent Coulomb peaks as high as $\sim e^2/h$. The B-dependence of the first two lower stripes reflects the ground-state evolution for $N = 3$ and 4. Their similar B-evolution indicates that the 3rd and 4th electron occupy the same orbital state with opposite spin, which is observed also for $N = 1$ and 2 (not shown). This is not the case for $N = 5$ and 6. The $N = 5$ state has $S = 1/2$, and the corresponding stripe shows a smooth evolution with B. Instead, the stripe for $N = 6$ has a kink at $B = B^* \approx 0.22\ T$. From earlier analyses [8] and from measurements of the excitation spectrum at finite bias V this kink is identified with a transition in the ground state from a spin-triplet to a spin-singlet.

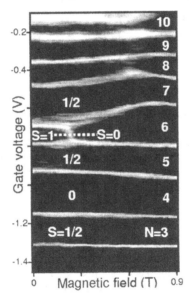

Fig. 6. Gray-scale representation of the linear conductance G versus the gate voltage V_g and the magnetic field B. White stripes denote conductance peaks of height $\sim e^2/h$. Dark regions of low conductance indicate Coulomb blockade. The $N = 6$ ground state undergoes a triplet-to-singlet transition at $B = B^* \approx 0.22\ T$, which results in a conductance anomaly inside the corresponding Coulomb gap.

Strikingly, at the triplet-singlet transition (see Fig. 6) we observe a strong enhancement of the conductance. In fact, over a narrow range around $0.22\ T$, the Coulomb gap for $N = 6$ has disappeared completely. Note that the change in greyscale along the dashed line in Fig. 6 represents the variation of the conductance with the tuning parameter K, see Fig. 3.

To explore this conductance anomaly, Fig. 7(a) shows the differential conductance, dI/dV versus V, taken at B and V_g corresponding to the intersection of the dotted line and the bright stripe $(B = B^*)$ in Fig. 6. The height of the zero-bias resonance decreases logarithmically with T [see Fig. 7(b)]. These are typical fingerprints of the Kondo effect. From FWHM≈ 30 $\mu V \approx k_B T_0$, we estimate $T_0 \approx 350$ mK. Note that $k_B T_0/g\mu_B B^* \approx 6$ so that the triplet state is indeed three-fold degenerate on the energy scale of T_0; this justifies an assumption made in Section 2.3 above. Also note that some of the traces in Fig. 7(a) show small short-period modulations which disappear above ~ 200 mK. These are due to a weak charging effect in the GaAs pillar above the dot [24].

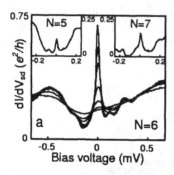 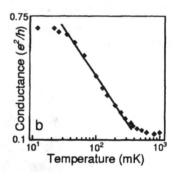

Fig. 7. (a) Kondo resonance at the singlet-triplet transition. The dI/dV vs V curves are taken at $V_g = -0.72$ V, $B = 0.21$ T and for $T = 14, 65, 100, 200, 350, 520$, and 810 mK. Kondo resonances for $N = 5$ (left inset) and $N = 7$ (right inset) are much weaker than for $N = 6$. (b) Peak height of zero-bias Kondo resonance vs T as obtained from (a). The line demonstrates a logarithmic T-dependence, which is characteristic for the Kondo effect. The saturation at low T is likely due to electronic noise.

For $N = 6$ the anomalous T-dependence is found only when the singlet and triplet states are degenerate. Away from the degeneracy, the valley conductance increases with T due to thermally activated transport. For $N = 5$ and 7, zero-bias resonances are clearly observed [see insets to Fig. 7(a)] which are related to the ordinary spin-1/2 Kondo effect. Their height, however, is much smaller than for the singlet-triplet Kondo effect.

We now investigate the effect of lifting the singlet-triplet degeneracy by changing B at a fixed V_g corresponding to the dotted line in Fig. 6. Near the edges of this line, i.e. away from B^*, the Coulomb gap is well developed as denoted by the dark colours. The dI/dV vs V traces still exhibit anomalies, however, now at finite V [see Fig. 8]. For $B = 0.21$ T we observe the singlet-triplet Kondo resonance at $V = 0$. At higher B this resonance splits apart showing two peaks at finite V, in agreement with the discussion above (see Section 2.3). For $B \approx 0.39$ T the peaks have evolved into steps which may indicate that the spin-coherence associated with the Kondo effect has completely vanished. The upper traces in Fig. 8, for $B < 0.21$ T, also show peak structures, although less pronounced.

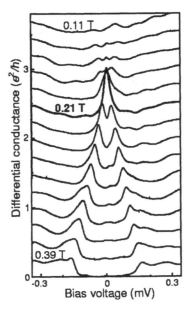

Fig. 8. dI/dV vs V characteristics taken along the dotted line in Fig. 6 at equally spaced magnetic fields $B = 0.11, 0.13, ..., 0.39\ T$. Curves are offset by $0.25\ e^2/h$.

3.2   Carbon Nanotubes

The situation in quantum dots formed in single-wall carbon nanotubes [25], [26],[15],[28] is rather different from that in semiconductor quantum dots. In nanotubes the effect of magnetic field on orbital motion is very weak, because the tube diameter ($\sim 1.4\ nm$) is an order of magnitude smaller than the magnetic length $l_B = (h/eB)^{1/2} \sim 10\ nm$ at a typical maximum laboratory field of $10\ T$. On the other hand, the g-factor is close to its bare value of $g = 2$, compared with $g = 0.44$ in GaAs. Hence the magnetic response of a nanotube dot is determined mainly by Zeeman shifts. As a result, the spins of levels in nanotube dots are easily measured [26],[28], and the ground state is usually (though not always [26]) found to alternate regularly between an $S = 0$ singlet for even electron number N and an $S = 1/2$ doublet for odd N [28],[29]. Moreover, singlet-triplet transitions in nanotubes are likely to be driven by the Zeeman splitting rather than orbital shifts, corresponding to the theory given in section 2.4.

   We discuss here the characteristics of a single-walled nanotube device with high contact transparencies, which were presented in more details in [29]. The source and drain contacts are gold, evaporated on top of laser–ablation–grown nanotubes [30] deposited on silicon dioxide. The conducting silicon substrate acts as the gate, as illustrated in Fig. 9. At room temperature the linear conductance G is $1.6\ e^2/h$, almost independent of gate voltage V_g, implying the conductance-dominating nanotube is metallic and defect free, and that the contact transmission coefficients are not much less than unity. At liquid helium temperatures regular Coulomb blockade oscillations develop, implying the formation of a single quantum dot limiting the conductance. However, the conductance in the Coulomb blockade valleys does not go to zero, consistent with high transmission coefficients and a strong coupling of electron states in the tube with the contacts.

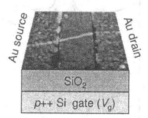

Fig. 9. Schematic of a nanotube quantum dot, incorporating an atomic force microscope image of a typical device (not the same one measured here.) Bridging the contacts, whose separation is 200 nm, is a 2 nm thick bundle of single-walled nanotubes

Fig. 10 shows a grayscale plot of dI/dV versus V and V_g over a small part of the full V_g range at $B = 0$. A regular series of faint "Coulomb diamonds" can be discerned, one of which is outlined by white dotted lines. Each diamond is labeled either E or O according to whether N is even or odd respectively, as determined from the effects of magnetic field. Superimposed on the diamonds are horizontal features which can be attributed to higher-order tunnelling processes that do not change the charge on the dot and therefore are not sensitive to V_g.

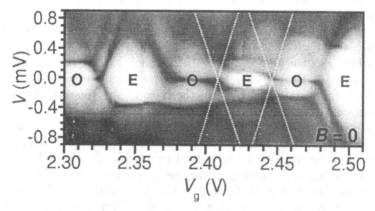

Fig. 10. Grayscale plot of differential conductance dI/dV (darker = more positive) against bias V and gate voltage V_g at a series of magnetic fields and base temperature (~ 75 mK). Labels 'E' and 'O' indicate whether the number of electrons N in the dot is even or odd (see text).

In Fig. 11(a) we concentrate on an adjacent pair of E and O diamonds in a magnetic field applied perpendicular to the tube. At $B = 0$ the diamond marked with an 'O' has narrow ridge of enhanced dI/dV spanning it at $V = 0$, while that marked with an 'E' does not. An appearence of a ridge at zero bias is consistent with formation of a Kondo resonance which occurs when N is odd (O) but not when it is even (E). This explanation is supported by the logarithmic temperature dependence of the linear conductance in the center of the ridge, as indicated in the inset to Fig. 12(b). At finite B, each zero-bias ridge splits into features at approximately $V = \pm E_Z/e$ as expected for Kondo resonances [4].

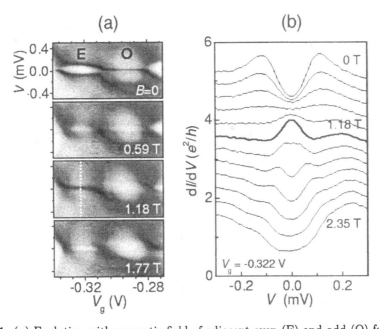

Fig. 11. (a) Evolution with magnetic field of adjacent even (E) and odd (O) features of the type seen in Fig. 10. (b) dI/dV vs V traces at the center of the E region, at $V_g = -0.322$ V. The trace at $B = B^* = 1.18$ T (bold line) corresponds to the dotted line in (a). The traces are offset from each other by $0.4\ e^2/h$ for clarity.

On the other hand, in the E diamond the horizontal features appear at a finite bias at $B = 0$. The origin of these features can be infered from their evolution with a magnetic field: while the ridge in the O region splits as B increases, the edges of the E 'bubble' move towards $V = 0$, finally merging into a single ridge at $B = B^* = 1.18$ T. Fig. 11(b) shows the evolution with B of the dI/dV vs V traces from the center of the E region, and the appearance of a zero-bias peak at around $B = 1.18$ T (bold trace). This matches what is expected for a Zeeman-driven singlet-triplet transition in the $N =$ even dots (Section 2.4). Further evidence that the peak is a Kondo resonance is provided by its temperature dependence [Fig. 12(a)], which shows an approximately logarithmic decrease of the peak height (the linear conductance) with T shown in Fig. 12(b).

Based on this interpretation we can deduce that for this particular value of N the energy gap separating the singlet ground state and the lowest-energy triplet state is $\Delta_0 = g\mu_B B^* \approx 137\ \mu eV$. At other even values of N the lowest visible excitations range in energy up to $\sim 400\ \mu eV$. For this device $E_C \sim 500\ \mu eV$. The energy gaps are therefore comparable with the expected single-particle level spacing δ, which is roughly equal to $E_C/3$ in a nanotube dot[15].

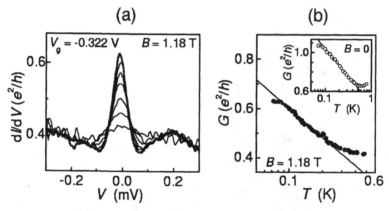

Fig. 12. (a) Temperature dependence at $B = B^*$. Here $T=$ base (bold line), 100, 115, 130, 180, 230, and 350 mK. (b) Temperature dependence of the linear conductance G (dI/dV at $V = 0$) at $B = B^* = 1.18\ T$. For comparison, $G(T)$ in the center of one of the O-type ridges at $B = 0$ is shown in the inset.

Note that the ridges at finite bias in Fig 11 in E valley are more visible at $B = 0$ than at $B = 0.59\ T$, halfway towards the degeneracy point. A possible explanation is that at $B = 0$ the triplet is not split, and all its components should be taken into account when calculating dI/dV at $B = 0$. This results in an enhancement of dI/dV at $B = 0$, $eV = \delta$, as compared to the value expected from the effective model of Section 2.4, which is valid in the vicinity of $B = B^*$.

Conclusion

Even a moderate magnetic field applied to a quantum dot or a segment of a nanotube can force a transition from the zero-spin ground state ($S = 0$) to a higher-spin state ($S = 1$ in our case). Therefore, the magnetic field may induce the Kondo effect in such a system. This is in contrast with the intuition developed on the conventional Kondo effect, which is destroyed by the applied magnetic field. In this paper we have reviewed the experimental and theoretical aspects of the recently studied magnetic–field–induced Kondo effect in quantum dots. Clearly there is more territory to be explored in the remarkably tuneable systems.

Acknowledgements

We thank our collaborators from Ben Gurion University, Delft University of Technology, Niels Bohr Institute, NTT, and University of Tokyo for their contributions. This work was supported by NSF under Grants DMR-9812340, DMR-9731756, by NEDO joint research program (NTDP-98), and by the EU via a TMR network. LG and MP are grateful to the Max Planck Institute for Physics of Complex Systems (Dresden, Germany), where a part of this paper was written, for the hospitality.

References

1. L. P. Kouwenhoven et al.: In: *Mesoscopic Electron Transport*, ed. by L. L. Sohn et al. (Kluwer, Dordrecht, 1997) pp. 105-214
2. C. B. Duke: *Tunneling in Solids* (New York, 1969); J. M. Rowell: In: *Tunneling Phenomena in Solids*, ed. by E. Burstein and S. Lundqvist (Plenum, New York, 1969)
3. A. F. G. Wyatt: Phys. Rev. Lett. **13**, 401 (1964); R. A. Logan and J. M. Rowell: Phys. Rev. Lett. **13**, 404 (1964)
4. J. Appelbaum: Phys. Rev. Lett. **17**, 91 (1966); Phys. Rev. **154**, 633 (1967); P. W. Anderson: Phys. Rev. Lett. **17**, 95 (1966)
5. J. Kondo: Prog. Theor. Phys. **32**, 37 (1964)
6. D. Goldhaber-Gordon et al.: Nature **391**, 156 (1998); S. M. Cronenwett, T. H. Oosterkamp, and L. P. Kouwenhoven, Science **281**, 540 (1998); J. Schmid et al.: Physica (Amsterdam) **256B-258B**, 182 (1998)
7. I. L. Kurland, I. L. Aleiner, and B. L. Altshuler: preprint cond-mat/0004205.
8. S. Tarucha et al.: Phys. Rev. Lett. **84**, 2485 (2000); D.G. Austing et al.: Phys. Rev. B **60**, 11514 (1999); L. P. Kouwenhoven et al.:, Science **278**, 1788 (1997)
9. M. Eto and Y. Nazarov: Phys. Rev. Lett. **85**, 1306 (2000)
10. M. Pustilnik and L. I. Glazman: Phys. Rev. Lett. **85**, 2993 (2000)
11. M. Pustilnik and L. I. Glazman: unpublished.
12. L. I. Glazman and M. E. Raikh: JETP Lett. **47**, 452 (1988); T. K. Ng and P. A. Lee: Phys. Rev. Lett. **61**, 1768 (1988)
13. T. Inoshita *et al.*: Phys. Rev. B **48**, 14725 (1993); L. I. Glazman, F. W. Hekking, and A. I. Larkin: Phys. Rev. Lett. **83**, 1830 (1999); A. Kaminski, L. I. Glazman: Phys. Rev. B **61**, 15297 (2000)
14. I. Affleck, A. W. W. Ludwig, and B. A. Jones: Phys. Rev. B **52**, 9528 (95); A. J. Millis, B. G. Kotliar, and B. A. Jones: In *Field Theories in Condensed Matter Physics*, ed. by Z. Tesanovic (Addison Wesley, Redwood City, CA, 1990), pp. 159-166
15. P. Nozières: J. Low Temp. Phys. **17**, 31 (1974)
16. P. W. Anderson: J. Phys. C **3**, 2436 (1970)
17. Y. Meir, N. S. Wingreen, and P. A. Lee: Phys. Rev. Lett. **70**, 2601 (1993); N. S. Wingreen and Y. Meir: Phys. Rev. B **49**, 11040 (1994)
18. A. Kaminski, Yu. V. Nazarov, and L. I. Glazman: Phys. Rev. Lett. **83**, 384 (1999); Phys. Rev. B **62**, 8154 (2000)
19. M. Pustilnik, Y. Avishai, and K. Kikoin: Phys. Rev. Lett. **84**, 1756 (2000)
20. I. E. Smolyarenko and N. S. Wingreen: Phys. Rev. B **60**, 9675 (1999)

21. A. M. Tsvelik and P. B. Wiegmann: Adv. Phys. **32**, 453 (1983)
22. T. A. Costi and A. C. Hewson, and V. Zlatić: J. Phys. CM **6**, 2519 (1994)
23. S. Sasaki et al.: Nature **405**, 764 (2000)
24. The top contact is obtained by deposition of Au/Ge and annealing at 400 °C for 30 s. This thermal treatment is gentle enough to prevent the formation of defects near the dot, but does not allow the complete suppression of the native Schottky barrier. The residual barrier leads to electronic confinement and corresponding charging effects in the GaAs pillar.
25. S. Tans et al.: Nature **386**, 474 (1997); C. Dekker, Physics Today **52**, 22 (1999)
26. S. Tans et al.: Nature **394**, 761 (1998)
27. M. Bockrath, et al.: Science **275**, 1922 (1997); M. Bockrath, et al.: Nature **397**, 598 (1999)
28. D. H. Cobden et al.: Phys. Rev. Lett. **81**, 681 (1998)
29. J. Nygård, D. H. Cobden, and P. E. Lindelof: Nature, *in press*
30. A. Thess et al.: Science **273**, 483 (1996)

Quantum Optics in Transport: Dicke Effect and Dark Resonances in Dots and Wires

Tobias Brandes

Universität Hamburg, 1. Institut für Theoretische Physik, Jungiusstr. 9, D-20355 Hamburg, Germany **

1 Introduction

The *Dicke effect* is known as spectral line narrowing due to collisions of atoms [1], and the Dicke superradiance [2], i.e. a collective decay of an ensemble of excited two–level systems due to spontaneous emission. Furthermore, *dark resonances* occur as quantum coherent 'trapped' superpositions of states in three–level systems under irradiation with two laser beams. Both effects are at the heart of mature and established branches in quantum optics and laser spectroscopy, existing and still growing over several decades with textbooks and numbers of review articles available [3–6].

There is a growing interest to transfer concepts from quantum optics and atomic physics to small semiconductor structures. The hope is that by combining methods like optical coherent control with the versatility of tunable mesoscopic systems such as quantum dots [7], one can realize artificial devices where basic quantum mechanical effects like superposition of states and quantum interference can be easily controlled.

In this contribution, we give a short survey of the Dicke effect in the transport properties of coupled quantum dots and quantum wires, and the coherent population trapping (dark resonance effect) in double quantum dots. The latter is a result of a collaboration with F. Renzoni. There are a number of related theoretical predictions by other authors that we cannot review here. Our intention is rather to introduce some of the basic physical concepts and ideas related to both effects in the new context of electronic transport; for more technical details we refer to the works cited in the text.

2 Dicke Superradiance in Electronic Systems

2.1 Sub– and Superradiance of Two Atoms

Let us start with a simple thought experiment: an excited atom, described as a two–level system (ground and excited state), decays due to spontaneous emission of photons. The decay rate Γ can be calculated from the matrix elements for

---

** Present and permanent address: Department of Physics, University of Manchester Institute of Science and Technology (UMIST), P.O. Box 88, Manchester M60 1QD, United Kingdom.

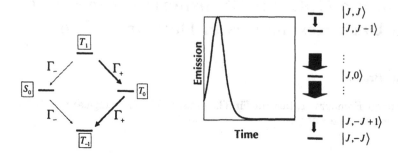

Fig. 1. Left: Decay scheme for a two–atom system. Indicated are the four states, Eq. (3), and the subradiant (Γ_-) and superradiant (Γ_+) decay channels. Right: Dicke peak in the emission of a superradiating ensemble of two–level systems as a function of time. The superradiant decay occurs by a down–cascade through the Dicke 'angular momentum' states $|JM\rangle$.

the interaction of the atom with the light. Considering now *two* (instead of one) atoms, this interaction (with the dipole moments $\hat{d}_1$ and $\hat{d}_2$ of the two atoms and g_Q the coupling matrix element) has the form

$$H_{eph} = \sum_Q g_Q \left(a_{-Q} + a_Q^+ \right) \left[\hat{d}_1 \exp i(\mathbf{Q}\mathbf{r}_1) + \hat{d}_2 \exp i(\mathbf{Q}\mathbf{r}_2) \right], \qquad (1)$$

from which it follows that the spontaneous emission rate Γ_Q of photons with wave vector $\mathbf{Q}$ is proportional to the square of the interaction (Fermi's Golden Rule),

$$\Gamma_\pm(Q) \propto \sum_Q |g_Q|^2 |1 \pm \exp\left[i\mathbf{Q}(\mathbf{r}_2 - \mathbf{r}_1)\right]|^2 \delta(\omega_0 - \omega_Q), \quad Q = \omega_0/c. \qquad (2)$$

Here, ω_0 is the transition frequency between the upper and lower level, and c denotes the speed of light. The interference of the two interaction contributions $\hat{d}_1 e^{i(\mathbf{Q}\mathbf{r}_1)}$ and $\hat{d}_2 e^{i(\mathbf{Q}\mathbf{r}_2)}$ leads to a splitting of the spontaneous decay into a fast, 'superradiant', decay channel ($\Gamma_+(Q)$), and a slow, 'subradiant' decay channel ($\Gamma_-(Q)$). This splitting is called 'Dicke–effect'.

Loosely speaking, the two signs $\pm$ correspond to the two different relative orientations of the dipole moments of the two atoms. More precisely, from the four possible states in the Hilbert space of two two–level systems, $\mathcal{H}_2 = C^2 \otimes C^2$, one can form singlet and triplet states according to

$$|S_0\rangle := \frac{1}{\sqrt{2}} \left(|\uparrow\downarrow\rangle - |\downarrow\uparrow\rangle \right)$$

$$|T_1\rangle := |\uparrow\uparrow\rangle, \quad |T_0\rangle := \frac{1}{\sqrt{2}} \left(|\uparrow\downarrow\rangle + |\downarrow\uparrow\rangle \right), \quad |T_{-1}\rangle := |\downarrow\downarrow\rangle. \qquad (3)$$

Fig. 1 indicates that the superradiant decay channel occurs via the triplet and the subradiant decay via the singlet states. In the extreme limit ('Dicke limit')

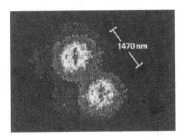 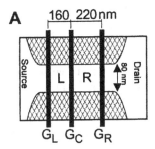

Fig. 2. Left: Image of a two–ion 'molecule' from the experiment of DeVoe and Brewer [8]. Right: Double quantum dots as used in the experiment by Fujisawa *et al.* [9] (top view). Transport of electrons is through the narrow channel that connects source and drain. The two quantum dots contain approximately 15 (Left, L) and 25 (Right, R) electrons. The charging energies are 4 meV (L) and 1 meV (R), the energy spacing for single particle states in both dots is approximately 0.5 meV (L) and 0.25 meV (R).

where the second phase factor is close to unity,

$$\exp\left[i\mathbf{Q}(\mathbf{r}_2 - \mathbf{r}_1)\right] \approx 1 \rightarrow \Gamma_-(Q) = 0, \quad \Gamma_+(Q) = 2\Gamma(Q), \qquad (4)$$

where $\Gamma(Q)$ is the decay rate of one *single* atom. This limit is achieved theoretically if $|\mathbf{Q}(\mathbf{r}_2 - \mathbf{r}_1)| \ll 1$ for all wave vectors $\mathbf{Q}$, i.e. the distance of the two atoms is much smaller than the wave length of the light. In practice, this 'pure' limit, where the subradiant rate is zero and the superradiant rate is just twice the rate for an individual atom, is never reached. It provides us, however, with a clear picture of what interference means for the process of spontaneous emission discussed here.

Fig. 2 (left) shows a recent experimental realization of sub– and superradiance from two laser–trapped ions. DeVoe and Brewer [8] measured the spontaneous emission rate of photons as a function of the ion–ion distance in a laser trap of planar geometry which was strong enough to bring the ions (Ba_{138}^+) to a distance of the order of $1\mu m$ of each other.

2.2   Spontaneous Emission of Phonons

Recently, the emission of *phonons* from double quantum dots (Fig. 2, right) has been observed [9]. Here, the coupling to the *phonon* degrees of freedom turned out to dominate the non–linear electron transport even at mK temperatures.

Basically, this system is a realization of an 'open' spin–1/2–boson [10] Hamiltonian, where 'open' means that due to a coupling to electron reservoirs, particles (electrons) can flow in and out of the system. In the strong Coulomb blockade regime, one additional electron can be either in the left or in the right dot. This corresponds to one single atom ($N = 1$) described by a 'pseudo' spin–1/2 with

z–component operator J_z. The interaction of this spin with the phonons,

$$H_{eph} = J_z \sum_{\mathbf{Q}} g_{\mathbf{Q}} \left[1 - \exp\left[i\mathbf{Q}(\mathbf{r}_L - \mathbf{r}_R)\right]\right] \left(a_{-\mathbf{Q}} + a_{\mathbf{Q}}^+\right), \qquad (5)$$

shows interference because the additional electron on the double dot can be either left or right [1]. This interference has the same form as in the subradiant coupling of photons to two atoms. It finally appears in the calculated inelastic current as a function of the energy difference ε between left and right dot [11,12], similar to what is observed in the experiments [9].

Double dots are controllable two–level systems offering the possibility to study the spin–boson dynamics in a well–defined, artificial environment. The question of what bosonic degrees of freedom are relevant can be addressed experimentally by, e.g., placing the sample into phonon cavities. Phonons become confined in free-standing structures, and this confinement can in principle be made visible in the electronic current [13]. Furthermore, one could imagine to 'scale up' interference effects (like the one discussed above) to a larger number of dots. The *collective* emission of phonons from $N > 1$ double quantum dots leads us to the Dicke superradiance effect.

2.3   N Atoms and the Dicke Effect

N atoms at positions $\mathbf{r}_i$ with dipole moments $\hat{d}_i$ interact with light by

$$H_{eph} = \sum_{\mathbf{Q}} g_{\mathbf{Q}} \left(a_{-\mathbf{Q}} + a_{\mathbf{Q}}^+\right) \sum_{j=1}^{N} \hat{d}_j \exp i(\mathbf{Q}\mathbf{r}_j). \qquad (6)$$

What Dicke realized in the 1950ties was the following: assume there is a situation in which one can approximate *all* the phase factors $e^{i\mathbf{Q}\mathbf{r}_j}$ by unity. Such a case might be fulfilled within a certain limit if the maximal distance between any two atoms is still much less than a typical wave length [2]. Then, the coupling to the photon field does no longer depend on the individual coordinate of the atom but it is only through the *collective* coordinate $\sum_{j=1}^{N} \hat{d}_j$.

At this point, we have to introduce some algebra. The dipole moments $\hat{d}_j$ of the individual atoms are described by Pauli matrices

$$\sigma_{x,i} = \begin{pmatrix} 0 & 1 \\ 1 & 0 \end{pmatrix}_i.$$

---

[1] In Eq. (5), an idealized situation of an extremely sharply peaked electron density (either in the center of the left or the right dot) has been assumed, and all form factors due to a smearing of this density (both in lateral and quantum well direction) have been set unity.

[2] An example is the case where all atoms are contained within a container, the size of which is less than the wave length of the emitted light. In this case, there is also no Doppler broadening because the collisions of the atoms with the container walls 'wipe out' the Doppler shifts $\mathbf{k}\mathbf{v}$ from the motion of atoms with velocity $\mathbf{v}$. This effect, which is called the Dicke spectral narrowing effect, was in fact discovered by Dicke in 1953 [1]. We discuss it below.

Since only the sum over all atoms enters, everything can be expressed by the sum of Pauli matrices. Effectively, we are adding up many 'spin 1/2's to a large spin which is described by angular momentum operators

$$J_z, \quad J_x \pm i J_y =: J_\pm, \quad [J_z, J_\pm] = \pm J_\pm, \quad [J_+, J_-] = 2J_z. \tag{7}$$

The total Hamiltonian now is simply

$$H = \hbar\omega_0 J_z + A J_x + H_{ph}, \tag{8}$$

where A is a photon operator and H_{ph} the photon field. The spontaneous decay of a completely inverted system (all electrons are in the upper levels) is then, corresponding to the superradiant decay through the triplet states for the case of $N = 2$ atoms, a 'down cascade' (Fig. 1 right) through what corresponds to the triplet states for higher N. This light peak consists of photons with different wave vectors $\mathbf{Q}$. The mean number $N_\mathbf{Q}(t)$ of photons as a function of time can be calculated exactly in this 'small–sample limit' of superradiance, see Agarwal [14]. It is proportional to the square of the coupling matrix element $g_\mathbf{Q}$ in the interaction Hamiltonian [3] Eq. (6). The 'Dicke states'[4] are denoted as eigenstates $|JM\rangle$ of the total spin $J = N/2$ and J_z, they form a basis of a $2J+1$–dimensional subspace of the Hilbert space of N two–level systems, $\mathcal{H}_N = (C^2)^{\otimes N}$.

What are the experimental consequences? An initially excited ensemble of N two–level systems decays in the form of a very sudden peak on a short time scale $\sim 1/N$, with the peak of the photon emission being $\sim N^2$, see Fig. 1[5]. The effect has been studied intensively from the 1970ties in atomic optics. There is a wealth of physical concepts related to superradiance: it involves concepts like coherence, dephasing (due to, e.g., van der Waals interactions) and the notion of a quasi-classical limit. As a transient process, superradiance occurs only if the observation time scale t is shorter than a *dephasing time scale* T_2 of processes that destroy phase coherence, and longer than the time τ which photons need to escape from the optical active region where the effect occurs [3], such that recombination processes are unimportant. This is the picture of a *dissipative decay* of the initially excited system due to spontaneous emission. In this approach, the photon system itself is in its vacuum state throughout the

---

[3] For example, for the case of $N = 2$ atoms in the small sample limit, one finds [14]

$$N_\mathbf{Q}(t \to \infty) = \frac{2|g_\mathbf{Q}|^2(X^2 + 40\Gamma^2)}{(X^2 + 16\Gamma^2)(X^2 + 4\Gamma^2)}, \quad X = \omega_0 - |\mathbf{k}|c, \tag{9}$$

where Γ is the decay rate of one single atom.

[4] J is also called *cooperation number*.

[5] In experiments, the small sample limit is never reached exactly. Instead of the collective operators J_z and $J_\pm$, one rather has to introduce $\mathbf{Q}$–dependent operators $J_\pm(\mathbf{Q}) := \sum_{j=1}^{N} J_\pm \exp i(\mathbf{Q}\mathbf{r}_j)$. Initial excitation with radiation in the form of a plane wave with wave vector $\mathbf{Q}$ leads to a collective state $\propto J_+(\mathbf{Q})^p|J, -J\rangle$ for some p. Subsequent spontaneous emission of photons with wave vector $\mathbf{Q}$ then is collective, conserves J and decreases $M \to M - 1$, while emission of photons with wave vector $\mathbf{Q}' \neq \mathbf{Q}$ can change J.

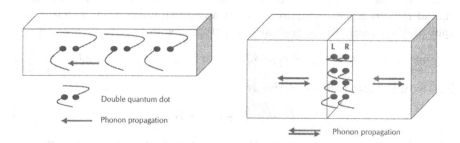

Fig. 3. Scheme for quasi one–dimensional superradiance in a slab (phonon–cavity) with several double quantum dots. The left case is a geometry suitable for one–mode superradiance with only one phonon wave vector $\mathbf{Q}$ supported by the structure. In this case, the double dots do not need to be on top of each other as in the right geometry.

time evolution. Any photon once emitted escapes from the system and there is no possibility of, e.g., re–absorption of photons. In particular this means that the collective radiation of the N atoms is *not* due to stimulated emission. In the spin–boson language, the back–action of the spin (which is described by the Schrödinger equation for its effective density matrix) onto the zero temperature bath of bosons is completely disregarded.

In a somewhat complementary approach, the dynamics of the photon field is treated on equal footing with the spin dynamics, and one has to solve the coupled 'Maxwell–Bloch' equations, i.e. the Heisenberg equations of motion in some decoupling approximation, of the total system (spin + photon field). This approach takes into account propagation effects of, e.g., an initial light pulse that excites the system, or re–absorption effects. The basic physics (i.e. the N^2 Dicke–peak), however, is the same in both pictures.

The condition

$$\tau \ll t \ll T_2, \Gamma^{-1}, \tag{10}$$

determines the superradiant regime [6], together with the last inequality which involves Γ^{-1}, the time scale for the decay of an *individual* atom. The restriction Eq. (24) of the *time–scale* for the superradiant process can be seen in analogy to the restriction

$$l \ll L \ll L_\phi \tag{11}$$

defining the *length scale* L of a mesoscopic system where physics occurs between a microscopic (e.g. atomic) length scale l and a dephasing length L_ϕ [15]. A superradiant system in some sense can be regarded as mesoscopic in time.

This wealth of physical concepts related to superradiance may in part have contributed to the quite recent revival of a considerable interest in the effect.

---

[6] For times t much larger than the dephasing time T_2, there is a transition to the regime of *amplified spontaneous emission* [4].

For example, coherent effects in semiconductors optics [16,17] have become accessible experimentally by ultrafast spectroscopy. The superradiance effect has been found in radiatively coupled quantum–well excitons [18–20]recently.

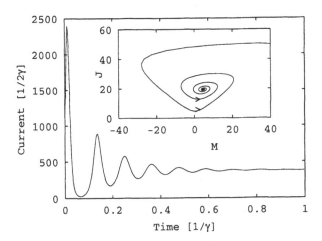

Fig. 4. Time–dependent inelastic current due to superradiance of phonons from a system of $N = 160$ double quantum dots. Inset: Plot in the space of Dicke states $|JM\rangle$ in the classical limit where J and M become continuous variables. Γ is the rate for tunneling of electrons to/from the leads, γ is the rate for inelastic transitions between left and right dots.

2.4 Superradiance in Coupled Double Dots

Fig. 3 schematically shows a system where superradiance can occur due to collective spontaneous emission of *phonons*. The model consists of an array of N double dots, interacting with phonons and coupled to electron reservoirs by leads. Electrons are pumped from the leads into the left dots, then tunnel inelastically to the right dots and finally enter the right reservoirs. This system can be shown [12] to behave like a large pseudo–spin which decays due to a *collective* spontaneous emission of phonons, if there is a large, identical bias $\varepsilon > 0$ between all left and all right dots. Furthermore, the 'reloading' with electrons into the left dots makes the spin to go up again and leads to an 'oscillatory superradiance' [21]. As a result, we have a transient behavior of both the current and the phonon emission rate in the form of an initial Dicke–peak, followed by oscillations up to a final stationary value, cf. Fig. 4.

3   Dicke Spectral Line Narrowing Effect

There are at least two effects in atomic physics named after Dicke. In his 1954 paper, Dicke coined the term **'superradiance'** for 'a gas which is radiating strongly because of coherence'. With his 1953 paper, he discovered that line shapes of absorption spectra of atoms in a gas can become very sharp due to velocity changing collisions (spectral line narrowing effect). Both effects are physically different from each other, but they are closely related from a more abstract (mathematical) point of view. Atomic spectral lines are broadened due

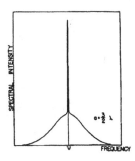

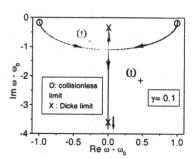

Fig. 5. Left: Line narrowing due to collisions of a Doppler–broadened spectral line in the original 1953 Dicke paper [1]. The radiating gas is modeled within a one-dimensional box of width a; λ is the light wavelength. Right: Zeros $\omega_\pm - \omega_0$ according to Eq.(13) appearing in the polarizability Eq.(12). The real and imaginary part of the frequencies are in units of the Doppler shift $v_p k$ which is fixed here. The two curves are plots parametric in the elastic collision rate $\Gamma(p)$; the arrows indicate the direction of increasing $\Gamma(p)$. For $\Gamma(p) \gg |v_p k|$, both curves approach the *Dicke limit* Eq.(14), where the imaginary part of $\omega_- - \omega_0$ becomes the negative of γ, and the imaginary part of $\omega_+ - \omega_0$ flows to minus infinity.

to Doppler frequency shifts $\mathbf{k v}$, where $\mathbf{v}$ is the velocity of an individual atom. Dicke showed that velocity–changing collisions of the radiating atoms with the atoms of a (non–radiating) buffer gas can average out this Doppler broadening. The result are spectral line shapes in the form of a very sharp peak on top of a broad line shape, centered around the transition frequency ω_0 of the atom, cf. Fig. 5.

3.1   Response Functions

The above effect in fact shows up quite generally in linear response functions under certain conditions. The relevant response function for spectral line shapes is the polarizability $\chi(k,\omega)$ of a gas (ensemble of moving two level systems) [22], where k is the wave vector and ω the frequency of an external electric field $E(k,\omega)$,

$$P(k,\omega) = \chi(k,\omega)E(k,\omega)$$
$$\chi(k,\omega) = d^2 \int dp N(p) \frac{\omega - \omega_0 + v_p k + i\gamma + 2i\Gamma(p)}{[\omega - \omega_+(p,k)][\omega - \omega_-(p,k)]} - (\omega_0 \to -\omega_0). \quad (12)$$

Here, d is the dipole moment, ω_0 the transition frequency, $N(p)$ the momentum distribution, $v_p = p/M$ with M the particle mass, γ is the natural line width, and $\Gamma(p)$ is the rate for elastic collisions at the atoms of a buffer gas that act like 'static impurities'. We have adopted a one–dimensional model for which the *poles* $\omega_\pm$ can be obtained as

$$\omega_\pm(p,k) := \omega_0 - i\gamma - i\left(\Gamma(p) \pm \sqrt{\Gamma(p)^2 - v_p^2 k^2}\right). \tag{13}$$

The poles in fact are eigenvalues of a collision matrix which, for the simplest case of only two poles, belong to symmetric and antisymmetric eigenmodes. As a function of an external parameter (e.g. the pressure of an atomic gas and therewith the collision rate), these poles can move through the lower frequency half–plane, whereby the spectral line shape becomes a superposition of a strongly broadened and a strongly sharpened peak. This phenomenon is in analogy with the formation of a bonding and an anti–bonding state by a coherent coupling of two (real) energy levels (level repulsion), with the difference that the Dicke effect is the splitting of decay rates, i.e. imaginary energies, into a fast (superradiant) and a slow (subradiant) mode. The position of the poles in the complex plane is shown in Fig.(5). We can distinguish two limiting cases:

In the collision-less limit $\Gamma^2(p) \ll v_p^2 k^2$, $\omega_\pm(p,k) = \omega_0 \pm v_p k - i\gamma$. The line-width is determined by the broadening through spontaneous emission γ and is shifted from the central position ω_0 by the *Doppler–shifts* $\pm v_p k$. Note that the final result for the polarizability still involves an integration over the distribution function $N(p)$ and therefore depends on the occupations of the upper and lower levels. This leads to the final *Doppler broadening* due to the Doppler–shifts $\pm v_p k$.

In the *Dicke–limit* $\Gamma^2(p) \gg v_p^2 k^2$, the Doppler-broadening can be neglected and

$$\omega_+ = \omega_0 - i\gamma - 2i\Gamma(p), \quad \omega_- = \omega_0 - i\gamma. \tag{14}$$

The first pole ω_+ corresponds to a broad resonance of width $\gamma + 2\Gamma(p)$, the second pole ω_- corresponds to a resonance whose width is solely determined by the 'natural' line–width γ, i.e. a resonance which is no longer Doppler–broadened.

The splitting into two qualitatively different decay channels is the key feature of the Dicke effect. We have already encountered it in the emission of light from a two–ion system, where the spontaneous decay split into one fast (superradiant) and one slow (subradiant) channel.

3.2 Spectral Function for Two Impurity Levels (Shahbazyan, Raikh)

The next example for this splitting is the model by Shahbazyan and Raikh [23] who considered resonant tunneling through two localized impurity levels of energy ε_1 and ε_2 (now we are back again in semiconductor physics!) The *conductance* of such a system can be expressed by its scattering properties, if

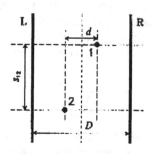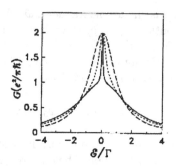

Fig. 6. Resonant tunneling through two impurity levels, from Shahbazyan and Raikh [23]. Left: Tunnel junction with two resonant impurities 1 and 2 in a distance d in horizontal and distance s_{12} in vertical direction. Left: linear conductance for identical impurity levels E as a function of $\mathcal{E} = E_F - E$, where E_F is the Fermi energy of the tunneling electron. The characteristic shape of the spectral function Eq.(18), as known from the Dicke effect, appears here in the conductance with increasing parameter $q = 0$, $q = 0.75$, $q = 0.95$. Γ is the tunneling rate through the left and the right barrier.

Coulomb interactions among the electrons are neglected [24–26]. The spectral function of an electronic system can be related to the imaginary part of its retarded Green's function [27,28]. For the case of two energy levels ε_1 and ε_2 that are assumed to belong to two spatially separated localized impurity states, one can define the spectral function in the Hilbert space of the two localized states where it becomes a two–by–two matrix,

$$S(\omega) = -\frac{1}{\pi}\Im\mathrm{m}\frac{1}{2}\mathrm{Tr}\frac{1}{\omega - \hat{\varepsilon} + i\hat{W}}. \tag{15}$$

It is assumed that transitions between the localized states $i \to |\mathbf{k}\rangle \to j$ are possible via virtual transitions to extended states (plane waves $|\mathbf{k}\rangle$). Such a situation is realized, e.g., for a coupling of two resonant impurity levels which are both coupled via a tunnel barrier to a continuum. The two levels then become coupled indirectly by virtual transitions of electrons from the impurities to the continuum and back again. Then, ε is diagonal in the ε_i, and $\hat{W}$ is a self–energy operator that describes the possibility of transitions between localized levels i and j via extended states with wave vector $\mathbf{k}$. In second order perturbation theory, the self–energy operator $\hat{W}$ is given by

$$W_{ij} = \pi \sum_{\mathbf{k}} t_{i\mathbf{k}} t_{\mathbf{k}j} \delta(\omega - E_{\mathbf{k}}), \tag{16}$$

where $\hbar = 1$ and the dependence on ω of $\hat{W}$ is no longer indicated. The quantities $t_{i\mathbf{k}}$ are overlaps between the localized states i and the plane waves $|\mathbf{k}\rangle$, their dependence on the impurity position $\mathbf{r}_i$ is given by the phase factor from the plane wave at the position of the impurity, i.e. $t_{i\mathbf{k}} \propto \exp(i\mathbf{k}\mathbf{r}_i)$. If identical diagonal elements $W_{11} = W_{22} = W$ are assumed for simplicity, the self–energy matrix becomes

$$i\hat{W} = iW \begin{pmatrix} 1 & q \\ q & 1 \end{pmatrix}, \quad q = J_0(r_{12}k_F), \tag{17}$$

and inversion of Eq.(15) yields

$$S(\omega) = \frac{1}{2\pi} \left[\frac{W_-}{(\omega - \varepsilon)^2 + W_-^2} + \frac{W_+}{(\omega - \varepsilon)^2 + W_+^2} \right], \quad W_\pm = (1 \pm q)W. \tag{18}$$

The eigenvalues $W_\pm$ of $\hat{W}$ determine the shape of the spectral function, which represents a superposition of two Lorentzians, i.e. one narrow line with width W_- and one broad line with width W_+. For distances r_{12} much less than the Fermi wave length, q is close to unity and the eigenvalues are 0 (subradiant channel) and $2W$ (superradiant channel).

3.3 The Dicke Effect in the AC Drude Conductivity of a Wire

Everything so far was, mathematically spoken, in the end related to a two–by–two matrix of the type

$$\hat{A} = \begin{pmatrix} \pm 1 & 1 \\ 1 & \pm 1 \end{pmatrix}, \tag{19}$$

which has eigenvalues 0 (subradiance) and ± 2 (superradiance). It is obvious that there will be many other physically interesting situations where the splitting of two poles within correlation functions plays a central role. The example discussed here is the AC conductivity of a disordered two–channel quantum wire in a magnetic field. The system and the subband dispersion relation are shown

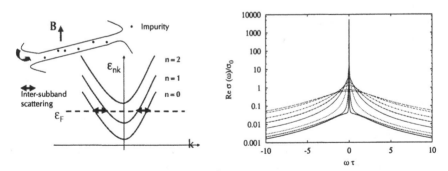

Fig. 7. Left: Subband dispersion ε_{nk} of a wire (upper left) in a magnetic field B. Right: Real part of the frequency-dependent Drude conductivity of a two–channel quantum wire in a magnetic field in units of $\sigma_0 := e^2 s v_{F_0} \tau / \pi$ (s=1 for spin–polarized electrons). The Dicke peak appears with increasing $\omega_c/\omega_0 = 0, 0.4, 0.8, 1.2, 1.6, 2.0, 2.4, 2.8$, where ω_0 is the frequency of the harmonic confinement potential, and $\omega_c = eB/m$ the cyclotron frequency for magnetic field B.

in Fig. 7. Many electronic transport properties of quantum wires have to be discussed in terms of the Landauer–Büttiker *conductance* Γ (the inverse resistance)

[24,25]. In some cases, a description in a Boltzmann–like theory by the Drude conductivity $\sigma(z)$ is meaningful in order to describe, e.g., an incoherent regime of transport. In a multichannel quasi one–dimensional wire, the AC Drude conductivity then can be expressed in terms of a scattering rate matrix $\hat{L}$ that describes scattering (forward and backward) between the different subbands,

$$\sigma(z) = ie^2 \sum_{nm} \left(\chi^0 [z\chi^0 + i\hat{L}]^{-1} \chi^0 \right)_{nm}. \tag{20}$$

If only two subbands of the wire are occupied, the matrix elements of $\hat{L}$ are

$$L_{00} = \frac{s}{\pi} L_s \frac{v_0}{v_1} \left(|V_{01}(k_0 - k_1)|^2 + |V_{01}(k_0 + k_1)|^2 + \frac{2s}{\pi} V_{00}(2k_0)^2 \right)$$

$$L_{11} = \frac{s}{\pi} L_s \frac{v_1}{v_0} \left(|V_{01}(k_0 - k_1)|^2 + |V_{01}(k_0 + k_1)|^2 + \frac{2s}{\pi} V_{11}(2k_1)^2 \right)$$

$$L_{01} = \frac{s}{\pi} L_s \left(|V_{01}(k_0 + k_1)|^2 - |V_{01}(k_0 - k_1)|^2 \right). \tag{21}$$

Furthermore, $\chi^0_{nm} = \delta_{nm} s v_n / \pi$, $s = 2$ ($s = 1$) if the electrons are spin degenerate (polarized), L_s is the length of the wire, v_i (k_i) are the Fermi velocities (wave vectors) in the two subbands $i = 0, 1$, and $|V_{ij}(k)|^2$ is the impurity averaged square of the scattering potential matrix element. The point now is that $\hat{L}$ for strong magnetic fields becomes, up to factors v_0/v_1, proportional to our 'Dicke–effect' matrix, i.e.

$$\hat{L} \propto \begin{pmatrix} -1 & 1 \\ 1 & -1 \end{pmatrix}. \tag{22}$$

The physical reason is that in strong magnetic fields, backscattering becomes suppressed and only the inter–subband forward scattering matrix elements $V_{01}(k_0 - k_1)$ survive. As a consequence, the real part of $\sigma(\omega)$ develops a stronger and stronger Dicke–peak, if the magnetic field is turned up. This is shown in Fig. 7 for a model with identical delta–scatterers of strength V_0^2, distributed with a concentration n_i^{2D} per area L^2, where a scattering rate $\tau^{-1} = n_i^{2D} V_0^2 m^* / \sqrt{4\pi} \hbar^3$ without magnetic field has been defined and the Fermi energy has been fixed between the bands $n = 1$ and $n = 2$, i.e. at $\varepsilon_F = 2\hbar\omega_B$, where $\omega_B := \sqrt{\omega_0^2 + \omega_c^2}$. Here, ω_0 is the confinement and ω_c the cyclotron frequency.

The effect should be observable in microwave absorption experiments. One should be able to vary ω such that $0.1 \lesssim \omega\tau \lesssim 5$ in order to scan the characteristic shape of the Dicke peak. Impurity scattering times for AlGaAs/GaAs heterostructures are between $3.8 \cdot 10^{-12}$ s and $3.8 \cdot 10^{-10}$ s for mobilities between $10^5 - 10^7$ cm^2/Vs, cp. [29]. A scattering time of 10^{-11} s requires frequencies of $\omega \approx 100$ GHz for $\omega\tau \approx 1$, which is consistent with the requirement of ω being much smaller than the effective confinement frequency ($\omega_0 = 1500$ Ghz for $\hbar\omega_0 = 1$ meV). The calculation applies for the case where the two lowest subbands are occupied. Temperatures T should be much lower than the subband–distance energy $\hbar\omega_B$, because thermal excitation of carriers would smear the effect. For $\hbar\omega_B$ of the order of a few meV, T should be of the order of a few

Kelvin or less. The Dicke peak appears for magnetic fields of a few Tesla such that ω_c/ω_0 becomes of the order and larger than unity.

4 Dark Resonances in Real and Artificial Atoms

4.1 Introduction

Quantum mechanical systems are described by state vectors. A linear superposition of two such vectors can result into a new state with physical properties that completely differ from the physical properties of the two original states. One ex-

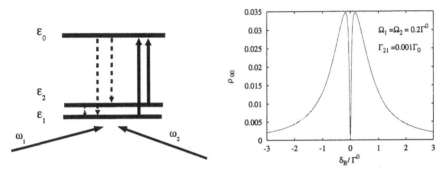

Fig. 8. Left: Three–level system under irradiation. Dashed lines indicate decay due to spontaneous emission of photons. Right: Stationary occupation of the upper level $|0\rangle$ of a three–level system under irradiation with two monochromatic light beams. Ω_1 and Ω_2 denote the Rabi frequencies corresponding to both beams, Γ^0 is the rate for spontaneous emission of photons from the upper level, and Γ_{21} is the rate for spontaneous emission of photons from level $|2\rangle$.

ample of such a superposition are the so–called **dark states** in atoms [6]. They are created by irradiating atoms with light of two frequencies ω_1 and ω_2. These frequencies can be tuned; they are chosen to be close to the transition frequencies for an electron that can jump between different levels of the atom. Usually, the dark states occur in three–level systems involving energy levels $\varepsilon_0 \gg \varepsilon_2 > \varepsilon_1$, see Fig. 8. There are other possible level configurations, but let us concentrate on this special choice. The transitions that are driven by the external light then are between the upper level ε_0 and ε_1, and between ε_0 and ε_2. By using the term 'transition' we actually have adopted a single–particle picture as is often done in atomic physics where the radiating properties of the atom are described by a single electron, hopping between different levels.

There is still one important ingredient missing: beside the external light (which only 'pumps' the electron within one of the two level pairs, i.e. $0 - 1$ or $0 - 2$), one needs a mechanism to shuffle the electron between 1 and 2. The dark state in fact is a superposition of the two states $|1\rangle$ and $|2\rangle$ belonging to these two levels. The shuffle mechanism is indeed a quite simple and very natural

one: it is provided by the decay of the upper level 0 into the two lower levels 1
and 2 by spontaneous emission of photons.

The states $|1\rangle$ and $|2\rangle$ are eigenstates of the Hamiltonian

$$H_0 := \sum_{i=0}^{2} \varepsilon_i |i\rangle\langle i| \tag{23}$$

of the atom without the coupling to electromagnetic fields. The external light
fields introduce an extra term to this Hamiltonian which then becomes

$$H(t) = H_0 + H_I(t), \quad H_I(t) = -\frac{\hbar\Omega_1}{2} e^{-i\omega_1 t}|0\rangle\langle 1| - \frac{\hbar\Omega_2}{2} e^{-i\omega_2 t}|0\rangle\langle 2| + h.c.. \tag{24}$$

Here, h.c. denoted the conjugate complex, and the light coupling is described
in the rotating wave approximation, where non–resonant terms corresponding to
very large energy transfer have been neglected. Furthermore, $\Omega_j = (E_j/\hbar)\langle 0|ez|j\rangle$,
$j = 1, 2$, are the Rabi frequencies, where E_j is the projection of the electric field
vectors of the light onto the dipole moments for the transitions $1 \rightarrow 0$, $2 \rightarrow 0$.

4.2 Dark States

Let us consider a simple case where the two levels 1 and 2 are degenerate, i.e.
$\varepsilon_1 = \varepsilon_2$, and furthermore assume that both frequencies $\omega_1 = \omega_2 = \omega$ and Rabi
frequencies $\Omega_1 = \Omega_2 = \Omega$ are identical. What we note immediately is that
the original states $|1\rangle$ and $|2\rangle$ are no longer eigenstates of the Hamiltonian $H(t)$
(there are no eigenstates of a time–dependent Hamiltonian anyway). Let us look,
however, at the superpositions

$$|\Psi_+\rangle = \frac{1}{\sqrt{2}} (|1\rangle + |2\rangle), \quad |\Psi_-\rangle = \frac{1}{\sqrt{2}} (|1\rangle - |2\rangle). \tag{25}$$

They are just a different choice of basis states for the two lowest levels 1 and
2. The 'minus'–state $|\Psi_-\rangle$, however, has the peculiar property that it remains
constant in the course of the time evolution, because

$$i\hbar\frac{d}{dt}|\Psi_-(t)\rangle = H(t)|\Psi_-(t)\rangle = 0, \tag{26}$$

as we can check from our Hamiltonian Eq. (24). Thus, the 'minus'–state is a
'dead–end': if the state of the atom evolves into $|\Psi_-\rangle$, it never gets out of this
state again. In particular, this state is 'dark' because the external light $H_I(t)$
cannot excite it to the upper state $|0\rangle$ (from which it could decay again and emit
photons).

On the other hand, the symmetric 'plus' state $|\Psi_+\rangle$ does not remain constant
in the course of the time evolution. In this state, the electron can absorb energy
from the external radiation and be transfered to the state $|0\rangle$.

We now also recognize the 'shuffle effect' of the decay from the upper level
$|0\rangle$: pumping of electrons up into $|0\rangle$ is only from the plus state $|\Psi_+\rangle$, whereas

spontaneous emission brings electrons to *both* $|\Psi_+\rangle$ and $|\Psi_-\rangle$ (which are superpositions of $|1\rangle$ and $|2\rangle$). The combination of this one–sided pumping with the two–fold decay eventually drives the atom into the 'dead–end' dark state $|\Psi_-\rangle$. It is somehow a funny fact that it is the decay mechanism (which is an inelastic process) that drives the atom into the superposition of two states. But don't forget that this is a superposition enforced by two time–dependent external fields.

4.3 Density Matrix

So far we have talked about quantum mechanical 'states'. It is more appropriate, however, to describe the three–level system by a statistical operator $\hat{\rho}(t)$, i.e. a density matrix that allows, e.g., a description of the spontaneous decay process. We do not want to go into the details of the equations of motion for this density matrix but only have a look at the parameters that enter, see table 1.

$\varepsilon_0 \gg \varepsilon_2 > \varepsilon_1$	energy levels of atom
ω_1, ω_2	angular frequencies of external light fields
Ω_1, Ω_2	Rabi frequencies of external light fields
$\delta_1 := \varepsilon_0 - \varepsilon_1 - \omega_1$	detuning off excitation energy $1 \to 0$
$\delta_2 := \varepsilon_0 - \varepsilon_2 - \omega_2$	detuning off excitation energy $2 \to 0$
$\delta_R := \delta_2 - \delta_1$	Raman detuning
$\Gamma_0/2$	Spontaneous decay rate from $|0\rangle$ to $|1\rangle$ and $|2\rangle$
Γ_{21}	Spontaneous decay rate from $|2\rangle$ to $|1\rangle$

Table 1. Parameters for coherent population trapping in three–level systems

Let us have a look at the results for the stationary values of the density matrix elements $\hat{\rho}_{ij}$. 'Stationary' means that (up to a trivial phase factor) these values do no longer change as a function of time. The best insight into the phenomenon comes from plotting the matrix elements as a function of the 'Raman' detuning δ_R, i.e. the difference of the relative detunings of the external light frequencies from the two transition frequencies. This can be achieved, e.g., by fixing the second frequency ω_2 exactly on resonance such that $\delta_2 = 0$ and varying $\omega_1 = -\delta_R$. The population $\hat{\rho}_{00}$ of the upper level then shows a typical resonance shape, i.e. it increases coming from large $|\delta_R|$ towards the center $\delta_R = 0$. Shortly before the resonance condition for the first light source, i.e. $\delta_R = 0$, is reached, the population drastically decreases in the form of a very sharp anti–resonance, up to a vanishing $\hat{\rho}_{00}$ for $\delta_R = 0$. This is the coherent population trapping effect. For $\delta_R = 0$, all the population (i.e. all the electrons in the ensemble of three–level systems) are trapped in the dark superposition $|\Psi-\rangle$ that can not be brought back to the excited state $|0\rangle$.

The dark state is a superposition of ground states that do not decay further, i.e. $|\Psi_-\rangle$ is stable. There is, however, a small probability for a decay from $|2\rangle$ to $|1\rangle$ as expressed by the rate Γ_{21} in table 1. This rate, together with the intensity of the external light, determines the small half-width $\delta_{1/2}$ of the antiresonance,

$$\delta_{1/2} \approx \frac{\Gamma_{21}}{2} + \frac{|\Omega_R|^2}{2\Gamma^0}, \tag{27}$$

where $\Omega_R := (\Omega_1^2 + \Omega_2^2)^{1/2}$.

4.4 Experiments, Applications, and Transfer to Double Quantum Dots

The first observation of dark states was in 1976 by Alzetta et al. [30] in the flourescence of sodium vapor. The effect, which was first considered as a curiosity, was quite soon recognized to be extremely useful for metrology, laser cooling and the adiabatic transfer of populations in molecular systems [6,31]. In the latter case, a coherence between the states 0 and 1 is established such that the state 2 becomes dark. The latter then is slowly shifted into the dark state 1 by changing the ratio of the two Rabi frequencies adiabatically, which is performed by applying two subsequent light pulses. Recently, investigations started with the aim to transfer such concepts to quantum wells [32] and optical experiments in quantum dots [33,34].

In the following, we describe how one can exploit the dark resonance effect to manipulate an electrical current, i.e. a transport property, of a double quantum dot. Ideally, one might like to find features like extremely sharp anti–resonances in a current as a function of an external parameter. Then, by changing the external parameter only a little bit, the current would drop from a finite value to zero. Such a 'current switch' could work if we were able to put electrical contacts to the levels of our three–level system. If one then could design a contact scheme such that the electrical current were proportional to the population of the upper level $|0\rangle$, it should work as an 'open' atom with current flowing through it and the light working as a mechanism to pump it into the dark, i.e. non–conducting state.

The realization of coherent optical effects in electronic systems, e.g. in artificial semiconductor structures, usually brings a lot of problems related to the vast amount of additional, unwanted processes like the incoherent interaction among the many electrons of such systems. In contrast to atoms or molecules (which are relatively simple systems), interactions in metals or semiconductors lead to what is called dephasing and make it in general difficult, e.g., to construct coherent superpositions of quantum states.

Fortunately, there has been an enormous technical progress in the production of semiconductor quantum dots which are, naively spoken, 'artificial atoms' with leads attached to them. The flow of electrons through such artificial atoms basically is governed by two physical mechanisms: the first one is the tunnel effect, i.e. electrons have to tunnel between the leads and the dots and between the dots, if two or more dots are contacted 'in series'. Second, the Coulomb blockade effect prevents too many electrons to pass through the dot at the same time. In the ideal case, electrons are transfered 'one–by–one' (single electron tunneling).

Our claim now is that *two* coupled quantum dots (so–called double dots) will do it for the dark resonance effect in the electrical current: The effect appears in double quantum dots where electron transport involves tunneling through two bonding and anti-bonding ground states $|1\rangle$ and $|2\rangle$ and one additional excited state $|0\rangle$, see Fig. 9.

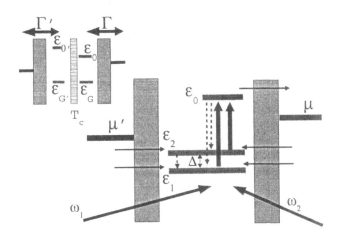

Fig. 9. Level scheme for two coupled quantum dots in Coulomb blockade regime. Two tunnel coupled ground states $|G\rangle$ and $|G'\rangle$ (small inset) form states $|1\rangle$ and $|2\rangle$ from which an electron is pumped to the excited state $|0\rangle$ by two light sources of frequency ω_1 and ω_2. Relaxation by acoustic phonon emission is indicated by dashed arrows. Γ and Γ' denote the rates for tunneling of electrons between the leads and the dots.

The main trick is to couple leads to both dots in such a way that electrons can tunnel into the ground states but leave the dot only through the excited state. This can be achieved by choosing an appropriate density of electrons in the leads, i.e. by tuning the chemical potentials of the leads. In addition, the system is irradiated with two light (microwave) sources with frequencies ω_1 and ω_2 that are detuned off the two excitation energies by $\hbar\delta_1 := \varepsilon_0 - \varepsilon_1 - \hbar\omega_1$ and $\hbar\delta_2 := \varepsilon_0 - \varepsilon_2 - \hbar\omega_2$. Relaxation from the excited level by acoustic phonon emission traps the dot in a coherent superposition of the bonding and the anti-bonding state, if $\delta_R := \delta_2 - \delta_1$ is tuned to zero. In this case, the excited level becomes completely depopulated.

Here we see the first important difference: in artificial atoms, acoustic phonon emission plays the role of the emission of photons from real atoms (or molecules). In the case of fluorescence emission of real atoms, the resulting trapping of the electron in a radiatively decoupled coherent superposition leads to 'dark resonances'. In the double dot case discussed here, the dark resonance effect appears as a suddenly vanishing electron current for $\delta_R = 0$. Although experiments have not yet been done, we suggest that for low enough microwave intensity, the effect can serve as a very sensitive, optically controlled current switch.

Atomic dark states have been found to be extraordinary stable against a number of perturbations [35]. In the quantum dot case, due to the Pauli blocking of the leads, a trapped electron can not tunnel out of the ground state coherent superposition. Furthermore, this superposition is protected from incoming electrons due to Coulomb blockade (no second electron can tunnel in). These two mechanisms guarantee the robustness of the effect, which is limited only by dephasing from inelastic processes. The latter are due to spontaneous emission of phonons in double dots [9,11] and can be controlled by tuning system parameters with gate voltages.

4.5 Predictions for Dark Resonances in Double Quantum Dots

We have made a theoretical analysis for the coherent population trapping effect in double quantum dots [36]. Again, one has to introduce a density matrix and to solve equations of motion (which we will not discuss in detail here). What is very important: the two rates Γ^0 and Γ_{21} for relaxation from state $|0\rangle$ and state $|2\rangle$ have to be calculated in order to estimate if or if not the effect can be seen in double dots at all. In fact, all of the story about the dark resonance here is based on the sharp antiresonance with a small half–width $\delta_{1/2}$. One can derive a formula for $\delta_{1/2}$ in the double dot case. For a symmetric configuration with $\varepsilon_G = \varepsilon_{G'}$, one finds

$$\delta_{1/2} \approx \frac{\Gamma_{21}}{2} + \frac{|\Omega_R|^2}{2[\Gamma^0 + \Gamma]}. \tag{28}$$

The relaxation rate Γ^0 depends on the energy ε_0 of the excited state $|0\rangle$ and is due to acoustic phonon emission. This rate depends on many other details of the system such as the size of the dots, the quantum well thickness and the material parameters for the electron–phonon coupling. We use parameters from known systems and obtain values of the order $\Gamma^0 \sim 10^{-9}$ s^{-1}.

The rate Γ_{21} is a bit more delicate: first, we must be very careful in its calculation since it determines the *minimal width* $\delta_{1/2}$ that we can achieve for very low light intensities, see Eq. (28). Here, we are in a fortunate situation because there are experiments which show that inelastic scattering processes due to tunneling between the two dots are related to the emission of piezoelectric phonons in GaAs [9,11]. With this knowledge at hand, one can even derive an analytical formula for Γ_{21}, namely

$$\Gamma_{21}(\Delta) \approx 2\pi \left(\frac{T_c}{\Delta}\right)^2 g \frac{\Delta}{\hbar} \left[1 - \frac{\sin(\Delta/\hbar\omega_d)}{\Delta/\hbar\omega_d}\right], \tag{29}$$

where T_c denotes the tunnel coupling matrix element between the two dots, $g \lesssim 0.05$ is the dimensionless phonon coupling constant, $\omega_d := c/d$ with c the longitudinal speed of sound and d the distance between the dot centers, and $\Delta := \varepsilon_2 - \varepsilon_1$.

Quantum dot experimentalists are able to tune this rate by tuning T_c (which in fact is possible by a skillful variation of gate voltages). This means they can in principle manipulate the quality of the coherent population trapping effect in their dots as they like! This is never possible in real atoms (unless one makes great efforts to suppress spontaneous emission of photons from a certain atomic level, if this is possible at all). Let us have a look how the theoretical predictions for the curves look like.

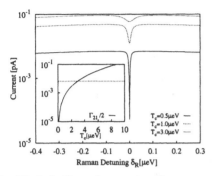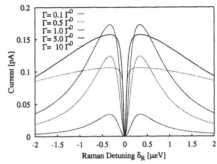

Fig. 10. Left: Tunnel current antiresonance through double dot system from Fig.9 with groundstate energy difference $\varepsilon = \varepsilon_{G'} - \varepsilon_G = 10\mu eV$. The Rabi frequencies Ω_1 and Ω_2 are taken to be equal, parameters are $\Omega_R = 0.2\Gamma^0$, $\Gamma = \Gamma' = \Gamma^0 = 10^9 s^{-1}$, where Γ^0 is the relaxation rate due to acoustic phonon emission from $|0\rangle$. Inset: Inelastic rate Γ_{21} (in $\mu eV/\hbar$), Eq. 29. Right: Current for fixed coupling T_c and different tunnel rates $\Gamma = \Gamma'$. Parameters are $\varepsilon = 10\mu eV$, $\Gamma^0 = 10^9 s^{-1}$, $\Omega_R = 1.0\Gamma^0$, $T_c = 1\mu eV$.

Fig. 10 shows the result for the stationary current I through the double dot as a function of the Raman detuning δ_R for the case with ground state energy difference $\varepsilon = \varepsilon_{G'} - \varepsilon_G = 10\mu eV$. We chose a fixed relaxation rate $\Gamma^0 = 10^9$ s^{-1}. Close to $\delta_R = 0$, the overall Lorentzian profile breaks in and shows a sharp current antiresonance. For fixed microwave intensity (fixed Rabi frequency $\Omega_R := (\Omega_1^2 + \Omega_2^2)^{1/2}$) and increasing tunnel coupling T_c, the inelastic rate Γ_{21}, Eq.(29), increases (inset). As a result, the antiresonance becomes broader and finally disappears for larger tunnel coupling T_c.

According to Eq. (28), $\delta_{1/2}$ increases with the inelastic rate Γ_{21}. For fixed microwave intensity, the vanishing of the antiresonance sets in for $\Gamma_{21} \gtrsim |\Omega_R|^2/[\Gamma^0 + \Gamma]$, cf. the inset of Fig. 10. On the other hand, with *increasing* elastic tunneling Γ out of the dot we recognize the striking fact that $\delta_{1/2}$ *decreases* down to its lower limit $\Gamma_{21}/2$. This behavior is shown in Fig. 10, right. For increasing tunnel rate Γ, the current increases until an overall maximal value is reached at $\Gamma \approx \Gamma^0$. The curve $I(\delta_R)$ decreases again and becomes very broad if the elastic tunneling becomes much faster than the inelastic relaxation Γ^0. Simultaneously, the center antiresonance then becomes sharper and sharper with increasing Γ, its half-width $\delta_{1/2}$ approaching the limit $\Gamma_{12}/2$, Eq. (28).

The resonance effect described above differs physically from other transport effects in AC–driven systems: coherent destruction of tunneling [37], tunneling through photo–sidebands [38], or coherent pumping of electrons [39,40] which are based on an additional time–dependent phase that electrons pick up while tunneling. In that case, the time evolution within the system is ideally completely coherent with dissipation being a disturbance rather then necessary for the effect to occur. In contrast, the trapping effect discussed here *requires* incoherent relaxation (phonon emission) within the system in order to create the trapped coherent superposition of the ground states.

5 Conclusion

It appears that the study of coherent effects in mesoscopic quantum transport is a very promising undertaking. In particular, the analogy between real and artificial atoms invites both theorists and experimentalists to transfer a plethora of various concepts and methods from quantum optics to mesoscopic semiconductor structures. The interaction of quantum dots with bosonic fields like phonons and photons offers a rich variety of effects that may allow to realize and to control basic quantum mechanical phenomena in artificial microscopic devices.

This work has been supported by the TMR network 'Quantum transport in the frequency and time domains', DFG projects Kr 627/9–1 and Br 1528/4–1, and the Graduiertenkolleg 'Physics of Nanostructures' at the University of Hamburg. The chapter on the coherent population trapping effect is the result of a collaboration with F. Renzoni, the chapter on Dicke superradiance was partly done in collaboration with J. Inoue and A. Shimizu. Discussions with R. Blick, S. Debald, B. Kramer, H. Schoeller, T. Vorrath, and W. van der Wiel are kindly acknowledged.

References

1. R. H. Dicke, Phys. Rev. **89**, 472 (1953).
2. R. H. Dicke, Phys. Rev. **93**, 99 (1954).
3. A. Andreev, V. Emel'yanov, and Y. A. Il'inski, *Cooperative Effects in Optics*, *Malvern Physics Series* (Institute of Physics, Bristol, 1993).
4. M. G. Benedict, A. M. Ermolaev, V. A. Malyshev, I. V. Sokolov, and E. D. Trifonov, *Super–Radiance, Optics and Optoelectronics Series* (Institute of Physics, Bristol, 1996).
5. M. Gross, S. Haroche, Phys. Rep. **93**, 301 (1982).
6. E. Arimondo, in *Progress in Optics*, edited by E. Wolf (Elsevier, Braunschweig, 1996), Vol. 35, p. 257.
7. L. Kouwenhoven and C. Marcus, Phys. World **June**, 35 (1998).
8. R. G. DeVoe and R. G. Brewer, Phys. Rev. Lett. **76**, 2049 (1996).
9. T. Fujisawa, T. H. Oosterkamp, W. G. van der Wiel, B. W. Broer, R. Aguado, S. Tarucha, and L. P. Kouwenhoven, Science **282**, 932 (1998); S. Tarucha, T. Fujisawa, K. Ono, D. G. Austin, T. H. Oosterkamp, and W. G. van der Wiel, Microelectr. Engineer. **47**, 101 (1999).
10. A. J. Leggett, S. Chakravarty, A. T. Dorsey and M. P. A. Fisher, A. Garg and W. Zwerger, Rev. Mod. Phys. **59**, 1 (1987).
11. T. Brandes and B. Kramer, Phys. Rev. Lett. **83**, 3021 (1999); Physica B **272**, 42 (1999); Physica B **284-288**, 1774 (2000) .

12. T. Brandes, Habilitation Thesis (University of Hamburg, Germany), 2000.
13. S. Debald, Master Thesis (University of Hamburg, Germany), 2000.
14. G. S. Agarwal, *Quantum Statistical Theories of Spontaneous Emission* (Springer, Berlin, 1974), Vol. 70.
15. Y. Imry, *Introduction to Mesoscopic Physics* (Oxford University Press, Oxford, 1997).
16. For a short recent review see, e.g., F. Rossi, Semicond. Sci. Technol. **13**, 147 (1998).
17. H. Haug and S. Koch, *Quantum Theory of the optical and electronic properties of semiconductors* (World Scientific, Singapore, 1990).
18. T. Stroucken, A. Knorr, P. Thomas, and S. W. Koch, Phys. Rev. B **53**, 2026 (1996).
19. S. Haas, T. Stroucken, M. Hübner, J. Kuhl, B. Grote, A. Knorr, F. Jahnke, S. W. Koch, R. Hey, K. Ploog, Phys. Rev. B **57**, 14860 (1998).
20. T. Stroucken, S. Haas, B. Grote, S. W. Koch, M. Hübner, D. Ammerlahn, J. Kuhl, in *Festkörperprobleme–Advances in Solid State Physics*, edited by B. Kramer (Vieweg, Braunschweig, 1998), Vol. 38, p. 265.
21. T. Brandes, J. Inoue, and A. Shimizu, Phys. Rev. Lett. **80**, 3952 (1998).
22. P. R. Berman, Appl. Phys. **6**, 283 (1975).
23. T. V. Shahbazyan and M. E. Raikh, Phys. Rev. B **49**, 17123 (1994).
24. R. Landauer, Philos. Mag. **21**, 863 (1970).
25. M. Büttiker, Phys. Rev. Lett. **57**, 1761 (1986).
26. H. Haug and A.-P. Jauho, in *Quantum Kinetics in Transport and Optics of Semiconductors*, Vol. 123 of *Solid-State Sciences* (Springer, Berlin, 1996), Chap. 12.
27. G. D. Mahan, *Many–Particle Physics* (Plenum Press, New York, 1990).
28. S. Doniach and E. H. Sondheimer, *Green's Functions for Solid State Physicists*, *Frontiers in Physics* (W. A. Benjamin, Reading, Massachusetts, 1974).
29. D. Ferry and S. Goodnik, *Transport in Nanostructures* (Cambridge University Press, Cambridge, UK, 1997).
30. G. Alzetta, A. Gozzini, L. Moi, G. Orriols, Nuovo Cimento B **36**, 5 (1976).
31. K. Bergmann, H. Theuer, B. W. Shore, Rev. Mod. Phys. **70**, 1003 (1998).
32. R. Binder, M. Lindberg, Phys. Rev. Lett. **81**, 1477 (1998).
33. N. H. Bonadeo, J. Erland, D. Gammon, D. Park, D. S. Katzer, and D. G. Steel, Science **282**, 1473 (1998).
34. U. Hohenester, F. Troiani, E. Molinari, G. Panzarini, and C. Macchiavello, Appl. Phys. Lett. **77**, 1864 (2000), and references therein.
35. M.C. de Lignie and E.R. Eliel, Opt. Comm. **72**, 205 (1989); E.R. Eliel, Adv. At. Mol Phys. **30**, 199 (1993).
36. T. Brandes, F. Renzoni, Phys. Rev. Lett. **85**, 4148 (2000).
37. F. Grossmann, T. Dittrich, P. Jung, and P. Hänggi, Phys. Rev. Lett. **67**, 516; M. Wagner, Phys. Rev. A **51**, 798 (1995).
38. C. Bruder and H. Schoeller, Phys. Rev. Lett. **72**, 1076 (1994); J. Iñarrea, G. Platero, and C. Tejedor, Phys. Rev. B **50**, 4581 (1994); Ph. Brune, C. Bruder, and H. Schoeller, Phys. Rev. B **56**, 4730 (1997).
39. C. A. Stafford and N. S. Wingreen, Phys. Rev. Lett. **76**, 1916 (1996).
40. M. Wagner and F. Sols, Phys. Rev. Lett. **83**, 4377 (1999).

Superconductors, Quantum Dots, and Spin Entanglement

Mahn-Soo Choi[1,2], C. Bruder[1], and Daniel Loss[1]

[1] Department of Physics and Astronomy, Klingelbergstrasse 82, CH-4056 Basel, Switzerland
[2] Korea Institute for Advanced Study, Cheongryangri-dong 207-43, Seoul 130-012, South Korea

Abstract. In this paper, we review a double quantum dot each dot of which is tunnel-coupled to superconducting leads. In the Coulomb blockade regime, a spin-dependent Josephson coupling between two superconductors is induced, as well as an antiferromagnetic Heisenberg exchange coupling between the spins on the double dot which can be tuned by the superconducting phase difference. We show that the correlated spin states—singlet or triplets—on the double dot can be probed via the Josephson current in a dc-SQUID setup. We also briefly review the Andreev entangler, a non-equilibrium setup that provides a source of pairwise entangled electrons.

1 Introduction

In recent years, electronic transport through strongly interacting mesoscopic systems has been the focus of many investigations [1]. In particular, a single quantum dot coupled via tunnel junctions to two non-interacting leads has provided a prototype model to study Coulomb blockade effects and resonant tunneling in such systems. These studies that started in the 1960's [2] have been extended to an Anderson impurity [3] or a quantum dot coupled to superconductors [4–6]. In a number of experimental [4] and theoretical [5] papers, the spectroscopic properties of a quantum dot coupled to two superconductors have been studied. Further, an effective dc Josephson effect through strongly interacting regions between superconducting leads has been analyzed [7–10]. More recently, on the other hand, research on the possibility to control and detect the spin of electrons through their charges has started. In particular in semiconducting nanostructures, it was found that the direct coupling of two quantum dots by a tunnel junction can be used to create entanglement between spins [11], and that such spin correlations can be observed in charge transport experiments [12].

Motivated by these studies we have proposed a new scenario for inducing and detecting spin correlations, viz., coupling a double quantum dot (DD) to superconducting leads by tunnel junctions [13]. It turns out that this connection via a superconductor induces a Heisenberg exchange coupling between the two spins on the DD. Moreover, if the DD is arranged between two superconductors, we obtain a Josephson junction (S-DD-S). The resulting Josephson current depends on the spin state of the DD and can be used to *probe* the spin correlations on the DD [13]. We have also pointed out that such a Josephson junction can be used in principle to distinguish singlet and triplet superconductors. Finally, a double quantum dot connected to a superconductor and to two leads has been proposed as an Andreev entangler, i.e., a device that allows the injection of spin-entangled electrons into two leads [14].

2   One Quantum Dot

As a warm-up, we would like to discuss a single quantum dot coupled to two superconducting leads (for a more detailed treatment see [3]). The geometry of the system is shown in Fig. 1.

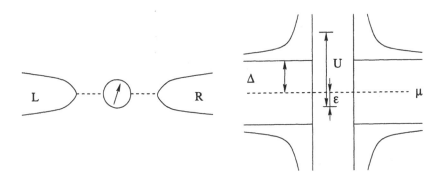

Fig. 1. Left panel: sketch of the superconductor-quantum dot-superconductor (S-D-S) nanostructure. Right panel: schematic representation of the quasiparticle energy spectrum in the superconductors and the energy levels of the quantum dot

The leads are assumed to be conventional singlet superconductors that are described by the BCS Hamiltonian

$$H_S = \sum_{j=L,R} \int_{\Omega_j} \frac{d\mathbf{r}}{\Omega_j} \left\{ \sum_{\sigma=\uparrow,\downarrow} \psi_\sigma^\dagger(\mathbf{r}) h(\mathbf{r}) \psi_\sigma(\mathbf{r}) + \Delta_j(\mathbf{r}) \psi_\uparrow^\dagger(\mathbf{r}) \psi_\downarrow^\dagger(\mathbf{r}) + h.c. \right\} , \quad (1)$$

where Ω_j is the volume of lead j, $h(\mathbf{r}) = (-i\hbar\nabla + \frac{e}{c}\mathbf{A})^2/2m - \mu$, and $\Delta_j(\mathbf{r}) = \Delta_j e^{-i\phi_j(\mathbf{r})}$ is the pair potential. For simplicity, we assume identical leads with same chemical potential μ, and $\Delta_L = \Delta_R = \Delta$. The quantum dot is modeled as a localized level ϵ with strong on-site Coulomb repulsion U, described by the Hamiltonian

$$H_D = -\epsilon \sum_\sigma d_{n\sigma}^\dagger d_{n\sigma} + U d_{n\uparrow}^\dagger d_{n\uparrow} d_{n\downarrow}^\dagger d_{n\downarrow} , \quad (2)$$

where $\epsilon > 0$. U is typically given by the charging energy of the dot, and we have assumed that the level spacing of the dot is $\sim U$ (which is the case for small GaAs dots [1]), so that we need to retain only one energy level in H_D. Finally, the dot is coupled (see Fig. 1) to the superconducting leads, described by the tunneling Hamiltonian

$$H_T = \sum_{j,\sigma} \left[t \exp\left(-i\frac{\pi}{\Phi_0} \int_{\mathbf{r}}^{\mathbf{r}_j} d\mathbf{l} \cdot \mathbf{A}\right) \psi_\sigma^\dagger(\mathbf{r}_j) d_\sigma + h.c. \right] , \tag{3}$$

where $\mathbf{r}_j$ is the point on the lead j closest to the dot. Here, $\Phi_0 = hc/2e$ is the superconducting flux quantum.

Since the low-energy states of the whole system are well separated by the superconducting gap Δ as well as the strong Coulomb repulsion U ($\Delta, \epsilon \ll U - \epsilon$), it is sufficient to consider an effective Hamiltonian on the reduced Hilbert space consisting of singly occupied levels of the dot and the BCS ground states on the leads. To lowest order in H_T, the effective Hamiltonian is given by [15]

$$H_{\text{eff}} = P H_T \left[(E_0 - H_0)^{-1} (1 - P) H_T \right]^3 P , \tag{4}$$

where P is the projection operator onto the subspace and E_0 is the ground-state energy of the unperturbed Hamiltonian H_0. (The second-order contribution leads to an irrelevant constant). The lowest-order expansion (4) is valid in the limit $\Gamma \ll \Delta, \epsilon$ where $\Gamma = \pi t^2 N(0)$ and $N(0)$ is the normal-state density of states per spin of the leads at the Fermi energy. Thus, we assume that $\Gamma \ll \Delta, \epsilon \ll U - \epsilon$, and temperatures which are less than ϵ (but larger than the Kondo temperature).

The explicit calculation of H_{eff} is a special case of the double-dot situation considered in the next section and is outlined in Appendix B. The result is

$$H_{\text{eff}} = \frac{J_0}{2} \cos(\phi) , \tag{5}$$

where ϕ is the phase difference between the two superconductors and J_0 is a *positive* constant given in Eq. (8). In other words, a quantum dot tunnel-coupled to two superconductors is a π-junction, the sign of the the Josephson coupling energy is opposite to that of a simple tunnel junction [2,3].

3   Two Quantum Dots

Now we would like to consider the double-dot (DD) system sketched in Fig. 2: Two quantum dots (a,b), each of which contains one (excess) electron and is connected to two superconducting leads (L, R) by tunnel junctions (indicated by dashed lines). Another realization would be atomic impurities embedded between the grains of a granular superconductor. There is no direct coupling between the two dots. The Hamiltonian describing this system consists of three parts, $H_S + H_{DD} + H_T \equiv H_0 + H_T$. The two quantum dots are modeled as two localized levels ϵ_a and ϵ_b with strong on-site Coulomb repulsion U, described by the Hamiltonian

$$H_D = \sum_{n=a,b} \left[-\epsilon \sum_\sigma d_{n\sigma}^\dagger d_{n\sigma} + U d_{n\uparrow}^\dagger d_{n\uparrow} d_{n\downarrow}^\dagger d_{n\downarrow} \right] , \tag{6}$$

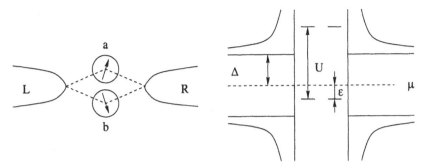

Fig. 2. Left panel: sketch of the superconductor-double quantum dot-superconductor (S-DD-S) nanostructure. Right panel: schematic representation of the quasiparticle energy spectrum in the superconductors and the single-electron levels of the two quantum dots

where we put $\epsilon_a = \epsilon_b = -\epsilon$ ($\epsilon > 0$) for simplicity. As before, we retain only one energy level per dot in H_{DD}. Finally, the DD is coupled *in parallel* (see Fig. 2) to the superconducting leads, described by the tunneling Hamiltonian

$$H_T = \sum_{j,n,\sigma} \left[t \exp(-i\frac{\pi}{\Phi_0} \int_{\mathbf{r}_n}^{\mathbf{r}_{j,n}} d\mathbf{l} \cdot \mathbf{A}) \psi_\sigma^\dagger(\mathbf{r}_{j,n}) d_{n\sigma} + h.c. \right] , \qquad (7)$$

where $\mathbf{r}_{j,n}$ is the point on the lead j closest to the dot n. Unless mentioned otherwise, it will be assumed that $\mathbf{r}_{L,a} = \mathbf{r}_{L,b} = \mathbf{r}_L$ and $\mathbf{r}_{R,a} = \mathbf{r}_{R,b} = \mathbf{r}_R$.

Proceeding as before, we calculate the effective Hamiltonian of the system to fourth order in H_T.

There are a number of virtual hopping processes that contribute to the effective Hamiltonian (4), see Fig. 3 for a partial listing and Fig. 5 for a full listing of them. Collecting these various processes, one can get the effective Hamiltonian in terms of the gauge-invariant phase differences ϕ and φ between the superconducting leads and the spin operators $\mathbf{S}_a$ and $\mathbf{S}_b$ of the dots (up to a constant and with $\hbar = 1$)

$$\begin{aligned} H_{\text{eff}} = {} & J_0 \cos(\pi f_{AB}) \cos(\phi - \pi f_{AB}) \\ & + [(2J_0 + J)(1 + \cos\varphi) + 2J_1(1 + \cos\pi f_{AB})] \left[\mathbf{S}_a \cdot \mathbf{S}_b - 1/4\right] . \end{aligned} \qquad (8)$$

Here $f_{AB} = \Phi_{AB}/\Phi_0$ and Φ_{AB} is the Aharonov-Bohm (AB) flux threading through the closed loop indicated by the dashed lines in Fig. 2. One should be careful to define *gauge-invariant* phase differences ϕ and φ in (8). The phase difference ϕ is defined as usual [16] by

$$\phi = \phi_L(\mathbf{r}_L) - \phi_R(\mathbf{r}_R) - \frac{2\pi}{\Phi_0} \int_{\mathbf{r}_R}^{\mathbf{r}_L} d\ell_a \cdot \mathbf{A} , \qquad (9)$$

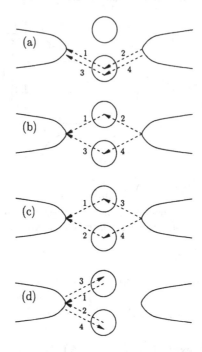

Fig. 3. Partial listing of virtual tunneling processes contributing to H_{eff}, Eq. (4). The numbered arrows indicate the direction and the order of occurrence of the charge transfers. Processes of type (a) and (b) give a contribution proportional to J_0, whereas those of type (c) and (d) give contributions proportional to J. For the complete list, see Fig. 5 in Appendix A

where the integration from $\mathbf{r}_R$ to $\mathbf{r}_L$ runs via dot a (see Fig. 2). The second phase difference, φ, is defined by

$$\varphi = \phi_L(\mathbf{r}_L) - \phi_R(\mathbf{r}_R) - \frac{\pi}{\Phi_0} \int_{\mathbf{r}_R}^{\mathbf{r}_L} (d\ell_a + d\ell_b) \cdot \mathbf{A} . \tag{10}$$

The distinction between ϕ and φ, however, is not significant unless one is interested in the effects of an AB flux through the closed loop in Fig. 2 (see Ref. [12] for an example of such effects). The coupling constants appearing in (8) are defined by

$$
\begin{aligned}
J &= \frac{2\Gamma^2}{\epsilon} \left[\frac{1}{\pi} \int \frac{dx}{f(x)g(x)} \right]^2 \\
J_0 &= \frac{\Gamma^2}{\Delta} \int \frac{dxdy}{\pi^2} \frac{1}{f(x)f(y)[f(x)+f(y)]g(x)g(y)} \\
J_1 &= \frac{\Gamma^2}{\Delta} \int \frac{dxdy}{\pi^2} \frac{g(x)[f(x)+f(y)]-2\zeta g(y)}{g(x)^2 g(y)[g(x)+g(y)][f(x)+f(y)]} ,
\end{aligned}
\tag{11}
$$

where $\zeta = \epsilon/\Delta$, $f(x) = \sqrt{1+x^2}$, and $g(x) = \sqrt{1+x^2}+\zeta$.

A remarkable feature of Eq. (8) is that a Heisenberg exchange coupling between the spin on dot a and on dot b is induced by the superconductor. This coupling is antiferromagnetic (all J's are positive) and thus favors a singlet ground state of spin a and b. This in turn is a direct consequence of the assumed singlet nature of the Cooper pairs in the superconductor (this is discussed further in the next section). As discussed below, an immediate observable consequence of H_{eff} is a *spin-dependent* Josephson current from the left to right superconducting lead (see Fig. 2) which probes the correlated spin state on the DD.

The various terms in (8) have different magnitudes. In particular, the processes leading to the J_1 term involve quasiparticles only as can be seen from its AB-flux dependence which has period $2\Phi_0$. In the limits we will consider below, this J_1 term is small and can be neglected.

In the limit $\zeta \gg 1$, the main contributions come from processes of the type depicted in Fig. 3 (a) and (b), making $J_0 \approx 0.1(\Gamma^2/\zeta\epsilon)\ln\zeta$ dominant over J and J_1. Thus, Eq. (8) can be reduced to

$$H_{\text{eff}} \approx J_0 \cos(\pi f_{AB}) \cos(\phi - \pi f_{AB}) + 2J_0(1 + \cos\varphi)\left[\mathbf{S}_a \cdot \mathbf{S}_b - \frac{1}{4}\right], \qquad (12)$$

up to order $(\ln\zeta)/\zeta$. As can be seen in Fig. 3 (a), the first term in Eq. (12) has the same origin as that in the single-dot case [3]: Each dot separately constitutes an effective Josephson junction with coupling energy $-J_0/2$ (i.e. π-junction) between the two superconductors. The two resulting junctions form a dc SQUID, leading to the total Josephson coupling in the first term of (12). The Josephson coupling in the second term in (12), corresponding to processes of type Fig. 3 (b), depends on the correlated spin states on the double dot: For the singlet state, it gives an ordinary Josephson junction with coupling $2J_0$ and competes with the first term, whereas it vanishes for the triplet states. Although the limit $\Delta \ll \epsilon \ll U - \epsilon$ is not easy to achieve with present-day technology, such a regime is relevant, say, for two atomic impurities embedded between the grains of a granular superconductor.

More interesting and experimentally feasible is the case $\zeta \ll 1$. In this regime, the effective Hamiltonian (8) is dominated by a single term (up to terms of order ζ),

$$H_{\text{eff}} \approx J(1 + \cos\varphi)\left[\mathbf{S}_a \cdot \mathbf{S}_b - \frac{1}{4}\right], \qquad (13)$$

with $J \approx 2\Gamma^2/\epsilon$. The processes of type Fig. 3 (b) and (c) give rise to (13). Below we will propose an experimental setup based on (13).

Before proceeding, we digress briefly on the dependence of J on the contact points. Unlike the processes of type Fig. 3 (a), those of types Fig. 3 (b), (c), and (d) depend on $\delta r_L = |\mathbf{r}_{L,a} - \mathbf{r}_{L,b}|$ and $\delta r_R = |\mathbf{r}_{R,a} - \mathbf{r}_{R,b}|$, see the remark below Eq. (7). For the tunneling Hamiltonian (7), one gets (putting $\delta r = \delta r_L = \delta r_R$)

$$J(\delta r) = \frac{8t^4}{\epsilon}\left|\int_0^\infty \frac{d\omega}{2\pi} \frac{F^R(\delta r, \omega) - F^A(\delta r, \omega)}{\omega + \epsilon}\right|^2, \qquad (14)$$

where $F^{R/A}(\mathbf{r}, \omega)$ is the Fourier transform of the Green's function in the superconductors, $F^{R/A}(\mathbf{r}, t) = \mp i\Theta(\pm t)\langle\{\psi_\uparrow(\mathbf{r}, t), \psi_\downarrow(0, 0)\}\rangle$. We note that the phase difference φ in (8) should now be defined with respect to the phase $\phi_j(\mathbf{r}_{j,a}, \mathbf{r}_{j,b})$ of the function $F_j^R(\mathbf{r}_{j,a}, \mathbf{r}_{j,b}) - F_j^A(\mathbf{r}_{j,a}, \mathbf{r}_{j,b})$ on the lead j, see the definition below (8). In the limit $\varepsilon \ll \Delta \ll \mu$, we find $J(\delta r) \approx J(0)e^{-2\delta r/\xi}\sin^2(k_F\delta r)/(k_F\delta r)^2$ up to order $1/k_F\xi$, with k_F the Fermi wave vector in the leads. Hence, the exchange coupling constant is exponentially suppressed if δr exceeds the superconducting coherence length ξ, and there is an additional suppression factor $1/(k_F\delta r)^2$.

4 Probing the Pairing Symmetry of the Superconducting Leads

In the previous section, the superconducting leads were assumed to be conventional BCS superconductors. The discovery of unconventional superconductivity in the heavy-fermion superconductor UPt_3 as well as the high-temperature superconductor $YBa_2Cu_3O_7$ has given a new impetus to the theoretical study of unconventional superconductors. These systems are characterized by an order parameter that is different from the symmetry of the underlying lattice. The order parameter has a nontrivial structure in k-space, usually accompanied by points or lines of zeroes in the gap. Also, the pairing symmetry in spin space that is of singlet type in conventional BCS superconductors can be of triplet type. This behavior is well-known from the p-wave triplet superfluid ^{3}He [17]. Recently, there has been strong evidence that Sr_2RuO_4 is a p-wave triplet superconductor [18].

If, in the previous section, we had assumed leads consisting of unconventional superconductors with triplet pairing, we would find a *ferromagnetic* exchange coupling favoring a triplet ground state of spin a and b on the DD. Thus, by probing the spin ground state of the dots (e.g. via its magnetic moment) we would have a means to *distinguish singlet from triplet pairing*. The magnetization could be made sufficiently large by extending the scheme from two to N dots or impurities coupled to the superconductor.

5 Probing Spins with a dc-SQUID

We now propose a possible experimental setup to probe the correlations (entanglement) of the spins on the dots, based on the effective model (13). According to (13) the S-DD-S structure can be regarded as a *spin-dependent* Josephson junction. Moreover, this structure can be connected with an ordinary Josephson junction to form a dc-SQUID-like geometry, see Fig. 4. The Hamiltonian of the entire system is then given by

$$H = J[1 + \cos(\theta - 2\pi f)]\left(\mathbf{S}_a \cdot \mathbf{S}_b - \frac{1}{4}\right) + \alpha J(1 - \cos\theta), \tag{15}$$

where $f = \Phi/\Phi_0$, Φ is the flux threading the SQUID loop, θ is the gauge-invariant phase difference across the auxiliary junction (J'), and $\alpha = J'/J$ with J' being the Josephson coupling energy of the auxiliary junction. Without restriction we can assume $\alpha > 1$, since J' could be adjusted accordingly by replacing the J'-junction by another dc SQUID and flux through it. One immediate consequence of Eq. (15) is that at zero temperature, we can effectively turn on and off the spin exchange interaction: For half-integer flux ($f = 1/2$), singlet and triplet states are degenerate at $\theta = 0$. Even at finite temperatures, where θ is subject to thermal fluctuations, singlet and triplet states are almost degenerate around $\theta = 0$. On the other hand, for integer flux ($f = 0$), the energy of the singlet state is lower by J than that of the triplet states.

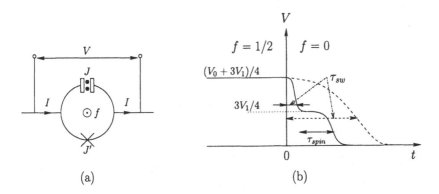

(a)   (b)

Fig. 4. (a) dc-SQUID-like geometry consisting of the S-DD-S structure (filled dots at the top) connected in parallel with another ordinary Josephson junction (cross at the bottom). (b) Schematic representation of dc voltage V vs. time when probing the spin correlations of the DD. The flux through the SQUID loop is switched from $f = 1/2$ to $f = 0$ at $t = 0$. Solid line: $\tau_{sw} < \tau_{spin}$. Dashed line: $\tau_{sw} > \tau_{spin}$

This observation allows us to probe directly the spin state on the double dot via a Josephson current across the dc-SQUID-like structure in Fig. 4. The supercurrent through the SQUID-ring is defined as $I_S = (2\pi c/\Phi_0)\partial\langle H\rangle/\partial\theta$, where the brackets refer to a spin expectation value on the DD. Thus, depending on the spin state on the DD we find

$$I_S/I_J = \begin{cases} \sin(\theta - 2\pi f) + \alpha\sin\theta & \text{(singlet)} \\ \alpha\sin\theta & \text{(triplets)} \end{cases}, \qquad (16)$$

where $I_J = 2eJ/\hbar$. When the system is biased by a dc current I larger than the spin- and flux-dependent critical current, given by $\max_\theta\{|I_S|\}$, a finite voltage V appears. Then one possible experimental procedure might be as follows (see Fig. 4b). Apply a dc bias current such that $\alpha I_J < I < (\alpha + 1)I_J$. Here, αI_J is the critical current of the triplet states, and $(\alpha + 1)I_J$ the critical current of the singlet state at $f = 0$, see (16). Initially prepare the system in an equal mixture of singlet and triplet states by tuning the flux around $f = 1/2$. (With electron g-factors $g \sim 0.5$–20 the Zeeman splitting on the dots is usually small compared with $k_B T$ and can thus be ignored.) The dc voltage measured in this mixture will be given by $(V_0 + 3V_1)/4$, where $V_0(V_1) \sim 2\Delta/e$ is the (current-dependent) voltage drop associated with the singlet (triplet) states. At a later time $t = 0$, the flux is switched off (i.e. $f = 0$), with I being kept fixed. The ensuing time evolution of the system is characterized by three time scales: the time $\tau_{coh} \sim \max\{1/\Delta, 1/\Gamma\} \sim 1/\Gamma$ it takes to establish coherence in the S-DD-S junction, the spin relaxation time τ_{spin} on the dot, and the switching time τ_{sw} to reach $f = 0$. We will assume $\tau_{coh} \ll \tau_{spin}, \tau_{sw}$, which is not unrealistic in view of measured spin decoherence times in GaAs exceeding 100ns [19]. If $\tau_{sw} < \tau_{spin}$, the voltage is given by $3V_1/4$ for times less than τ_{spin}, i.e. the singlet no longer contributes to the voltage. For $t > \tau_{spin}$ the spins have relaxed to their ground (singlet) state, and the voltage vanishes. One therefore expects steps in the voltage versus time (solid curve in Fig. 4b). If $\tau_{spin} < \tau_{sw}$, a broad transition region of the voltage from the initial value to 0 will occur (dashed line in Fig. 4b).

Another experimental setup would be to use an rf-SQUID geometry, i.e., to embed the S-DD-S structure into a superconducting ring [16]. However, to operate such a device, ac fields are necessary, and the sensitivity is not as good as for the dc-SQUID geometry.

To our knowledge, there are no experimental reports on quantum dots coupled to superconductors. However, hybrid systems consisting of superconductors (e.g., Al or Nb) and 2DES (InAs and GaAs) have been investigated by a number of groups [20]. Taking the parameters of those materials, a rough estimate leads to a coupling energy J in Eqs. (13) or (15) of about $J \sim 0.05$–0.5K. This corresponds to a critical current scale of $I_J \sim 5$–50nA.

6   Andreev Entangler

Up to now, we have considered equilibrium phenomena. Recently, a system consisting of a double quantum dot coupled to a superconductor on one side and to separate leads on the other side has been proposed as a source of entangled electrons [14]. This entangler is a *non-equilibrium* device that can create pairwise spin-entangled electrons and provide coherent injection by an Andreev process into different dots which are tunnel-coupled to leads, leading to a current

$$I_1 = \frac{4e\gamma_S^2}{\gamma} \left[\frac{\sin(k_F \delta r)}{k_F \delta r} \right]^2 \exp\left\{ -\frac{2\delta r}{\pi \xi} \right\}. \tag{17}$$

Here, γ_S is the tunneling rate between the superconductor and the dots, γ the tunneling rate between the dots and the leads, and δr was defined around Eq. (14).

The unwanted process of both electrons tunneling into the same leads can be suppressed by increasing the Coulomb repulsion on the quantum dot, and its current is given by

$$I_2 = \frac{2e\gamma_S^2\gamma}{\mathcal{E}^2}, \qquad \frac{1}{\mathcal{E}} = \frac{1}{\pi\Delta} + \frac{1}{U}. \tag{18}$$

These relations are valid if $\Delta, U > \Delta\mu > \gamma, k_B T$, and $\gamma > \gamma_S$, where $\Delta\mu$ is the bias voltage between the superconductor and the leads. Also, the single-particle level spacing of both dots is assumed to be larger than $\Delta\mu$.

The ratio of currents of these two competing processes is given by

$$\frac{I_1}{I_2} = \frac{2\mathcal{E}^2}{\gamma^2} \left[\frac{\sin(k_F \delta r)}{k_F \delta r} \right]^2 \exp\{ -\frac{2\delta r}{\pi \xi} \} . \tag{19}$$

From this ratio we see that the desired regime with I_1 dominating I_2 is obtained when $\mathcal{E}/\gamma > k_F \delta r$, and $\delta r < \xi$. We would like to emphasize that the relative suppression of I_2 (as well as the absolute value of the current I_1) is maximized by working around the resonances $\epsilon_l \simeq \mu_S = 0$. It was shown that there exists a regime of experimental interest where the entangled current shows a resonance and assumes a finite value with both partners of the singlet being in different leads but having the same orbital energy. This entangler then satisfies the necessary requirements needed to detect the spin entanglement via transport and noise measurements [12].

Another effect discussed in [14] are flux-dependent oscillations of the current in an Aharonov-Bohm loop. For this let us consider now a setup where the two leads 1 and 2 are connected such that they form an Aharonov-Bohm loop, where the electrons are injected from the left via the superconductor, traversing the upper (lead 1) and lower (lead 2) arm of the loop before they rejoin to interfere and then exit into the same lead, where the current is then measured as a function of varying flux Φ. It is straightforward to analyze this setup. The total flux-dependent Aharonov-Bohm current I_{AB} is found to be [14]

$$I_{AB} = \sqrt{8 I_1 I_2} F(\epsilon_l) \cos(\Phi/2\Phi_0) + I_2 \cos(\Phi/\Phi_0) , \tag{20}$$

$$F(\epsilon_l) = \frac{\epsilon_l}{\sqrt{\epsilon_l^2 + (\gamma_L/2)^2}} , \tag{21}$$

where, for simplicity, we have assumed that $\epsilon_1 = \epsilon_2 = \epsilon_l$, and $\gamma_1 = \gamma_2 = \gamma_L$. Here, the first term (different leads) is periodic in $2\Phi_0 = h/e$ like for single-electron Aharonov-Bohm interference effects, while the second one (same leads) is periodic in the superconducting flux quantum Φ_0, describing thus the interference of two coherent electrons (similar single- and two-particle Aharonov-Bohm effects occur in the Josephson current through an Aharonov-Bohm loop, see the previous sections and [13]). It is clear from Eq. (20) that the h/e oscillation comes from the interference between a contribution where the two electrons travel through different arms with contributions where the two electrons travel through the same arm. Both Aharonov-Bohm oscillations with period h/e, and $h/2e$, vanish with decreasing I_2, i.e. with increasing on-site repulsion U and/or gap Δ. However, their relative weight is given by $\sqrt{I_1/I_2}$, implying that the $h/2e$ oscillations vanish faster than the h/e ones. This behavior is quite remarkable since it opens up the possibility to tune down the unwanted leakage process $\sim I_2 \cos(\Phi/\Phi_0)$ where two electrons proceed via the same dot/lead by increasing U with a gate voltage applied to the dots. The dominant current contribution with period h/e comes then from the desired entangled electrons proceeding via different leads. On the other hand, if $\sqrt{I_1/I_2} < 1$, which could become the case e.g. for $k_F \delta r > \mathcal{E}/\gamma$, we are left with $h/2e$ oscillations only. Note that dephasing processes which affect the orbital part suppress I_{AB}. Still, the flux-independent current $I_1 + I_2$ can remain finite and contain electrons which are entangled in spin-space, provided that there is only negligible spin-orbit coupling so that the spin is still a good quantum number.

In conclusion, we have reviewed double quantum dots each dot of which is coupled to superconductors. We have found that in the Coulomb blockade regime the Josephson current from one superconducting lead to the other is different for singlet or triplet states on the double dot. This leads to the possibility to probe the spin states of the dot electrons by measuring a Josephson current. We have discussed the possibility to use a Josephson junction of this type to distinguish between singlet and triplet superconductors. And finally, we have briefly reviewed a non-equilibrium device: the recently proposed Andreev entangler, a source of entangled electrons.

We would like to thank the Swiss National Science Foundation for support.

Appendix A

In this appendix, we would like to enumerate the processes contributing to the effective Hamiltonian, Eq. (4). They are depicted in Fig. 5 and have labels A_1, B_1, B_2, ... E_2.

Each process can be calculated in a straightforward way; as examples, we give the explicit calculations for A_1 and C_1 in the following two appendices. Adding up all of these terms, we get

$$
\begin{aligned}
H_{\text{eff}} = {} & J_0 \cos(\pi f) \cos\phi & \text{Class } A_1 \\
& - Y_1 & \text{Class } B_1 \\
& - (X_1 + Y_2) & \text{Class } B_2 \\
& + J \cos\phi \, [\mathbf{S}_a \cdot \mathbf{S}_b - 1/4] & \text{Class } C_1 \\
& + 2J_0 \cos\phi \, [\mathbf{S}_a \cdot \mathbf{S}_b - 1/4] & \text{Class } C_2 \\
& - 2Y_1 \cos(\pi f) \, [\mathbf{S}_a \cdot \mathbf{S}_b + 1/4] & \text{Class } D_1 \\
& + 2(X_1 + Y_2) \cos(\pi f) \, [\mathbf{S}_a \cdot \mathbf{S}_b + 1/4] & \text{Class } D_2 \\
& + (J + 2X_1 + 2Y_2) \, [\mathbf{S}_a \cdot \mathbf{S}_b - 1/4] & \text{Class } E_1 \\
& + 2J_0[\mathbf{S}_a \cdot \mathbf{S}_b - 1/4] - 2Y_1[\mathbf{S}_a \cdot \mathbf{S}_b + 1/4] & \text{Class } E_2
\end{aligned}
$$

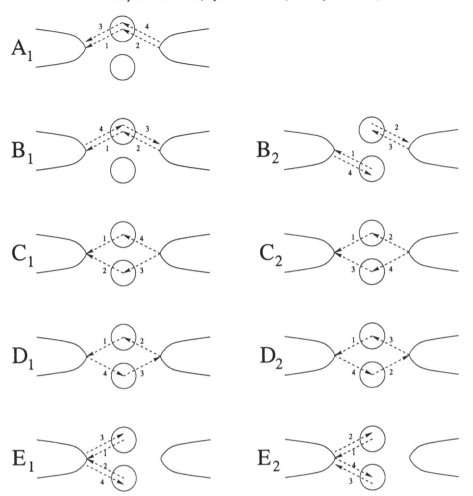

Fig. 5. Processes contributing to the effective Hamiltonian, Eq. (4)

which, after simplification, can be written in the following form:

$$
\begin{aligned}
H_{\text{eff}} = {} & J_0 \cos(\pi f) \cos(\varphi - \pi f) \\
& + 2J_0(1 + \cos\phi)\, [\mathbf{S}_a \cdot \mathbf{S}_b - 1/4] \\
& + 2J_1[1 + \cos(\pi f)]\, [\mathbf{S}_a \cdot \mathbf{S}_b - 1/4] \\
& + J(1 + \cos\phi)\, [\mathbf{S}_a \cdot \mathbf{S}_b - 1/4] \\
& + J_1[\cos(\pi f) - 1] - 3Y_1 \ .
\end{aligned}
\tag{22}
$$

The coupling constants in these expressions are given by

$$J_0 = \frac{\Gamma^2}{\Delta} \iint \frac{dxdy}{\pi^2} \frac{1}{f(x)f(y)[f(x)+f(y)]g(x)g(y)} \tag{23}$$

$$J_1 = X_1 - X_2 \tag{24}$$

$$J = \frac{2\Gamma^2}{\epsilon} \left[\frac{1}{\pi} \int \frac{dx}{f(x)g(x)} \right]^2 \tag{25}$$

$$X_1 = \frac{\Gamma^2}{\Delta} \iint \frac{dxdy}{\pi^2} \frac{1}{g(x)g(y)[g(x)+g(y)]} \tag{26}$$

$$X_2 = \frac{\Gamma^2}{\Delta} \iint \frac{dxdy}{\pi^2} \frac{2\epsilon/\Delta}{g^2(x)[f(x)+f(y)][g(x)+g(y)]} \equiv Y_1 - Y_2 \tag{27}$$

$$Y_1 = \frac{\Gamma^2}{\Delta} \iint \frac{dxdy}{\pi^2} \frac{1}{[f(x)+f(y)]g^2(x)} \tag{28}$$

$$Y_2 = \frac{\Gamma^2}{\Delta} \iint \frac{dxdy}{\pi^2} \frac{1}{g^2(x)[g(x)+g(y)]} \tag{29}$$

$$f(x) \equiv \sqrt{1+x^2}, \quad g(x) \equiv \sqrt{1+x^2} + \epsilon/\Delta. \tag{30}$$

They can be derived as follows:

$$J_0 = 4t^4 \sum_{k\in L} \sum_{q\in R} \frac{|u_k v_k^* u_q^* v_q|}{(E_k+\epsilon)(E_q+\epsilon)(E_k+E_q)}$$
$$= t^4 \sum_{kq} \frac{|\Delta_L \Delta_R|}{E_k E_q (E_k+E_q)(E_k+\epsilon)(E_q+\epsilon)} \tag{31}$$

$$J_1 = X_1 - X_2 \tag{32}$$

$$J = 8\frac{t^4}{\epsilon} \left| \sum_k \frac{u_k v_k^*}{E_k+\epsilon} \right|^2 = \frac{2t^4}{\epsilon} \left| \sum_k \frac{\Delta^*}{E_k(E_k+\epsilon)} \right|^2 \tag{33}$$

$$X_1 = 4t^4 \sum_{kq} \frac{|u_k|^2 |u_q|^2}{(E_k+\epsilon)(E_q+\epsilon)(E_k+E_q+2\epsilon)}$$
$$= t^4 \sum_{kq} \frac{(E_k+\xi_k)(E_q+\xi_q)}{E_k E_q (E_k+\epsilon)(E_q+\epsilon)(E_k+E_q+2\epsilon)} \tag{34}$$

$$X_2 = Y_1 - Y_2 = 4t^2 \sum_{kq} \frac{2\epsilon |u_k|^2 |v_k|^2}{(E_k+\epsilon)^2(E_k+E_q)(E_k+E_q+2\epsilon)} \tag{35}$$

$$Y_1 = 4t^4 \sum_{kp} \frac{|u_k|^2 |v_p|^2}{(E_k+\epsilon)^2(E_k+E_q)} = t^4 \sum_{kp} \frac{(E_k+\xi_k)(E_p-\xi_p)}{E_k E_p (E_k+E_q)(E_k+\epsilon)^2} \tag{36}$$

$$Y_2 = 4t^4 \sum_{kq} \frac{|u_k|^2 |u_q|^2}{(E_k + \epsilon)^2 (E_k + E_q + 2\epsilon)} = t^4 \sum_{kq} \frac{(E_k + \xi_k)(E_q + \xi_q)}{E_k E_q (E_k + \epsilon)^2 (E_k + E_q + 2\epsilon)} . \quad (37)$$

More explicitly, with $\zeta = \epsilon/\Delta$, we obtain

$$\frac{X_1}{|\Delta|} = \frac{t^4}{|\Delta|^4} [N(0)|\Delta|]^2$$

$$\times \int \frac{dxdy \, (\sqrt{1+x^2} + x)(\sqrt{1+y^2} + y)}{\sqrt{1+x^2}(\sqrt{1+x^2} + |\zeta|)\sqrt{1+y^2}(\sqrt{1+y^2} + |\zeta|)(\sqrt{1+x^2} + \sqrt{1+y^2} + |\zeta|)}$$

$$= \frac{t^4}{|\Delta|^4} [N(0)|\Delta|]^2 \int \frac{dxdy}{(\sqrt{1+x^2} + |\zeta|)(\sqrt{1+y^2} + |\zeta|)(\sqrt{1+x^2} + \sqrt{1+y^2} + 2|\zeta|)}$$

$$\qquad (38)$$

$$\frac{Y_2}{|\Delta|} = \frac{t^4}{|\Delta|^4} [N(0)|\Delta|]^2$$

$$\times \int \frac{dxdy \, (\sqrt{1+x^2} + x)(\sqrt{1+y^2} + y)}{\sqrt{1+x^2}(\sqrt{1+x^2} + |\zeta|)^2 \sqrt{1+y^2}(\sqrt{1+x^2} + \sqrt{1+y^2} + |\zeta|)} \quad (39)$$

$$= \frac{t^4}{|\Delta|^4} [N(0)|\Delta|]^2 \int \frac{dxdy}{(\sqrt{1+x^2} + |\zeta|)^2 (\sqrt{1+x^2} + \sqrt{1+y^2} + 2|\zeta|)} .$$

Appendix B

Here, we would like to evaluate explicitly the contribution of process A_1 (see Fig. 5) to the effective Hamiltonian.

Below, L and U denotes the lower and upper dot. The states that we consider have the form $|G; \sigma_a, \sigma_b; G\rangle$, where σ_a, σ_b denotes the spin states of the dots and G is the BCS ground state of the superconductors. In these calculations, we have retained the k-dependence of the tunneling matrix elements; they are denoted e.g. t_{ka} for the tunneling to dot a. In the end, we set $t_{ka} = t$.

$$|G; \uparrow, \downarrow; G\rangle = d_{a\uparrow}^\dagger d_{b\downarrow}^\dagger |G; 0, 0; G\rangle$$

$$= d_{a\uparrow}^\dagger d_{b\downarrow}^\dagger \prod_k \left(u_k + v_k c_{k\uparrow}^\dagger c_{-k\downarrow}^\dagger \right) \prod_q \left(u_q + v_q c_{q\uparrow}^\dagger c_{-q\downarrow}^\dagger \right) |0; 0, 0; 0\rangle$$

$$\xrightarrow{H_T} \sum_k t_{ka} u_k \gamma_{k0}^\dagger d_{a\uparrow} d_{a\uparrow}^\dagger d_{b\downarrow}^\dagger |G; 0, 0; G\rangle = \sum_k t_{ka} u_k \gamma_{k0}^\dagger |G; 0, \downarrow; G\rangle$$

$$\xrightarrow{\frac{1-P}{H_0}} + \sum_{k \in L} t_{ka} \frac{u_k}{E_k - \epsilon} \gamma_{k0}^\dagger |G; 0, \downarrow; G\rangle$$

$$\xrightarrow{H_T} + \sum_{q \in R} t_{qa}^* v_q d_{a\uparrow}^\dagger \gamma_{q1}^\dagger \sum_{k \in L} t_{ka} \frac{u_k}{E_k - \epsilon} \gamma_{k0}^\dagger |G; 0, \downarrow; G\rangle$$

$$- \sum_{q \in R} t_{qa}^* v_q d_{a\downarrow}^\dagger \gamma_{q0}^\dagger \sum_{k \in L} t_{ka} \frac{u_k}{E_k - \epsilon} \gamma_{k0}^\dagger |G; 0, \downarrow; G\rangle$$

$$= + \sum_{k \in L} \sum_{q \in R} t_{ka} t_{qa}^* \frac{u_k v_q}{E_k - \epsilon} \gamma_{q1}^\dagger \gamma_{k0}^\dagger d_{a\uparrow}^\dagger |G; 0, \downarrow; G\rangle$$

$$- \sum_{k \in L} \sum_{q \in R} t_{ka} t_{qa}^* \frac{u_k v_q}{E_k - \epsilon} \gamma_{q0}^\dagger \gamma_{k0}^\dagger d_{a\downarrow}^\dagger |G; 0, \downarrow; G\rangle$$

$$\xrightarrow{\frac{1-P}{H_0}} + \sum_{k \in L} \sum_{q \in R} t_{ka} t_{qa}^* \frac{u_k v_q}{(E_k - \epsilon)(E_k + E_q)} \gamma_{q1}^\dagger \gamma_{k0}^\dagger d_{a\uparrow}^\dagger |G; 0, \downarrow; G\rangle$$

$$- \sum_{k \in L} \sum_{q \in R} t_{ka} t_{qa}^* \frac{u_k v_q}{(E_k - \epsilon)(E_k + E_q)} \gamma_{q0}^\dagger \gamma_{k0}^\dagger d_{a\downarrow}^\dagger |G; 0, \downarrow; G\rangle$$

$$\xrightarrow{H_T} + \sum_{l \in L} t_{la} u_l \gamma_{l0}^\dagger d_{a\uparrow} \sum_{k \in L} \sum_{q \in R} t_{ka} t_{qa}^* \frac{u_k v_q}{(E_k - \epsilon)(E_k + E_q)} \gamma_{q1}^\dagger \gamma_{k0}^\dagger d_{a\uparrow}^\dagger |G; 0, \downarrow; G\rangle$$

$$+ \sum_{l \in L} t_{la} u_l \gamma_{l1}^\dagger d_{a\downarrow} (-1) \sum_{k \in L} \sum_{q \in R} t_{ka} t_{qa}^* \frac{u_k v_q}{(E_k - \epsilon)(E_k + E_q)} \gamma_{q0}^\dagger \gamma_{k0}^\dagger d_{a\downarrow}^\dagger |G; 0, \downarrow; G\rangle$$

$$- \sum_{l \in L} t_{la} v_l^* \gamma_{l0} d_{a\downarrow} (-1) \sum_{k \in L} \sum_{q \in R} t_{ka} t_{qa}^* \frac{u_k v_q}{(E_k - \epsilon)(E_k + E_q)} \gamma_{q0}^\dagger \gamma_{k0}^\dagger d_{a\downarrow}^\dagger |G; 0, \downarrow; G\rangle$$

$$= + \sum_{kl \in L} \sum_{q \in R} t_{ka} t_{la} t_{qa}^* \frac{u_k u_l v_q}{(E_k - \epsilon)(E_k + E_q)} \gamma_{l0}^\dagger \gamma_{q1}^\dagger \gamma_{k0}^\dagger |G; 0, \downarrow; G\rangle$$

$$- \sum_{kl \in L} \sum_{q \in R} t_{ka} t_{la} t_{qa}^* \frac{u_k u_l v_q}{(E_k - \epsilon)(E_k + E_q)} \gamma_{l1}^\dagger \gamma_{q0}^\dagger \gamma_{k0}^\dagger |G; 0, \downarrow; G\rangle$$

$$- \sum_{k \in L} \sum_{q \in R} t_{ka}^2 t_{qa}^* \frac{u_k v_k^* v_q}{(E_k - \epsilon)(E_k + E_q)} \gamma_{q0}^\dagger |G; 0, \downarrow; G\rangle .$$

Here the contributions of the first and second terms will vanish because one cannot get the final BCS ground state for the SC from these states.

$$\xrightarrow{\frac{1-P}{H_0}} - \sum_{k \in L} \sum_{q \in R} t_{ka}^2 t_{qa}^* \frac{u_k v_k^* v_q}{(E_k - \epsilon)(E_k + E_q)(E_q - \epsilon)} \gamma_{q0}^\dagger |G; 0, \downarrow; G\rangle$$

$$\xrightarrow{H_T} + t_{qa}^* u_q^* d_{a\uparrow}^\dagger \gamma_{q0} (-1) \sum_{k \in L} \sum_{q \in R} t_{ka}^2 t_{qa}^* \frac{u_k v_k^* v_q}{(E_k - \epsilon)(E_k + E_q)(E_q - \epsilon)} \gamma_{q0}^\dagger |G; 0, \downarrow; G\rangle$$

$$= - \sum_{k \in L} \sum_{q \in R} t_{ka} t_{ka} t_{qa}^* t_{qa}^* \frac{u_k v_k^* u_q^* v_q}{(E_k - \epsilon)(E_k + E_q)(E_q - \epsilon)} |G; \uparrow, \downarrow; G\rangle$$

$$= - t^4 e^{-i4\pi f/4} \sum_{k \in L} \sum_{q \in R} \frac{\Delta_L^* \Delta_R}{4 E_k (E_k - \epsilon) E_q (E_q - \epsilon)(E_k + E_q)} |G; \uparrow, \downarrow; G\rangle$$

$$= - \frac{1}{4} J_0 e^{i(\phi - 4\pi f/4)} |G; \uparrow, \downarrow; G\rangle .$$

The path which transfers Cooper pairs from the left to the right superconductor contributes the complex conjugate of the above one. On the other hand, the paths through the upper dots have contributions with factor $e^{+i4\pi f/4}$.

Here, it should be noticed that ϕ is *not* gauge invariant. Therefore it is useful to rewrite

$$t_{ka}t_{ka}t^*_{qa}t^*_{qa}\Delta^*_L \Delta_R = t^4 \exp\left(-i\,2 \times \frac{2\pi}{2\Phi_0}\int_R^L d\ell_a \cdot \mathbf{A}\right)\exp[i(\phi_L - \phi_R)] \tag{40}$$

$$= t^4 \exp\left[i\left(\phi_L - \phi_R - \frac{2\pi}{\Phi_0}\int_R^L d\ell_a \cdot \mathbf{A}\right)\right] \tag{41}$$

$$\equiv t^4 e^{+i\varphi_a}\,. \tag{42}$$

As the result of this calculation, we obtain

$$H^{(4)}_{A_1} = \frac{J_0}{2}\left[\cos\varphi_a + \cos\varphi_b\right] = J_0 \cos(\pi f)\cos(\varphi_b - \pi f)\,, \tag{43}$$

where

$$\theta_n \equiv \phi_L - \phi_R - \frac{2\pi}{\Phi_0}\int_R^L d\ell_n \cdot \mathbf{A}\,. \tag{44}$$

It is interesting to think about the effect discussed above in the following way: Class A_1 processes describe the fact that each quantum dot forms a π-junction between the two superconductors. We thus have two π-junctions linked in a loop as in a dc-SQUID, through which AB flux threads. The total energy for this configuration is given by

$$H^{(1)}_{A_1} = \frac{1}{2}J_0 \cos\varphi + \frac{1}{2}J_0 \cos(\varphi - 2\pi f) = J_0 \cos(\pi f)\cos(\varphi - \pi f)\,. \tag{45}$$

Appendix C

Here, we would like to evaluate explicitly the contribution of process C_1 to the effective Hamiltonian. The hopping events labeled 1 to 4 in Fig. 5 can be combined in a different order that have to be considered separately. We distinguish these different "paths" by labels P_i, $i = 1, 2, ...$ and consider the relevant possibilities.

$$\langle G; \uparrow, \downarrow; G| \, H^{(4)} \, |G; \uparrow, \downarrow; G\rangle \begin{cases} 4 \leftarrow 3 \leftarrow 2 \leftarrow 1 & : P_1 \\ 3 \leftarrow 4 \leftarrow 2 \leftarrow 1 & : P_2 \\ 4 \leftarrow 3 \leftarrow 1 \leftarrow 2 & : P_3 \\ 3 \leftarrow 4 \leftarrow 1 \leftarrow 2 & : P_4 \end{cases}$$

$$\langle G; \downarrow, \uparrow; G| \, H^{(4)} \, |G; \uparrow, \downarrow; G\rangle \begin{cases} 4 \leftarrow 3 \leftarrow 2 \leftarrow 1 & : P_5 \\ 3 \leftarrow 4 \leftarrow 2 \leftarrow 1 & : P_6 \\ 4 \leftarrow 3 \leftarrow 1 \leftarrow 2 & : P_7 \\ 3 \leftarrow 4 \leftarrow 1 \leftarrow 2 & : P_8 \end{cases}$$

Neither the configuration $|G;\uparrow,\uparrow;G\rangle$ nor $|G;\downarrow,\downarrow;G\rangle$ can gain energy via co-tunneling of this type.

Path P_1:

$$|G;\uparrow,\downarrow;G\rangle = d_{a\uparrow}^\dagger d_{b\downarrow}^\dagger |G;0,0;G\rangle$$

$$= d_{a\uparrow}^\dagger d_{b\downarrow}^\dagger \prod_k \left(u_k + v_k c_{k\uparrow}^\dagger c_{-k\downarrow}^\dagger\right) \prod_q \left(u_q + v_q c_{q\uparrow}^\dagger c_{-q\downarrow}^\dagger\right) |0;0,0;0\rangle$$

$$\xrightarrow{H_T} \sum_{k\in L} t_{ka} u_k \gamma_{k0}^\dagger d_{a\uparrow} \, d_{a\uparrow}^\dagger d_{b\downarrow}^\dagger |G;0,0;G\rangle = \sum_{k\in L} t_{ka} u_k \gamma_{k0}^\dagger d_{b\downarrow}^\dagger |G;0,0;G\rangle$$

$$\xrightarrow{\frac{1-P}{H_0}} \sum_{k\in L} t_{ka} \frac{u_k}{E_k - \epsilon} \gamma_{k0}^\dagger d_{b\downarrow}^\dagger |G;0,0;G\rangle$$

$$\xrightarrow{H_T} + \sum_{l\in L} t_{lb} u_l \gamma_{l1}^\dagger d_{b\downarrow} \sum_{k\in L} t_{ka} \frac{u_k}{E_k - \epsilon} \gamma_{k0}^\dagger d_{b\downarrow}^\dagger |G;0,0;G\rangle$$

$$+ \sum_{l\in L} t_{lb} v_l^* \gamma_{l0} d_{b\downarrow} \sum_{k\in L} t_{ka} \frac{u_k}{E_k - \epsilon} \gamma_{k0}^\dagger d_{b\downarrow}^\dagger |G;0,0;G\rangle$$

$$= + \sum_{kl\in L} t_{ka} t_{lb} \frac{u_k u_l}{E_k - \epsilon} \gamma_{k0}^\dagger \gamma_{l1}^\dagger |G;0,0;G\rangle$$

$$+ \sum_{k\in L} t_{kl} t_{lb} \frac{u_k v_k^*}{E_k - \epsilon} |G;0,0;G\rangle$$

$$\xrightarrow{\frac{1-P}{H_0}} + \sum_{kl\in L} t_{ka} t_{lb} \frac{u_k u_l}{(E_k - \epsilon)(E_k + E_l - 2\epsilon)} \gamma_{k0}^\dagger \gamma_{l1}^\dagger |G;0,0;G\rangle$$

$$+ \sum_{k\in L} t_{ka} t_{lb} \frac{u_k v_k^*}{(E_k - \epsilon)(-2\epsilon)} |G;0,0;G\rangle \ .$$

The first term will be projected out at the end.

$$\xrightarrow{H_T} + \sum_{p\in R} t_{pb}^* u_p^* d_{b\downarrow}^\dagger \gamma_{p1} \sum_{k\in L} t_{ka} t_{lb} \frac{u_k v_k^*}{(E_k - \epsilon)(-2\epsilon)} |G;0,0;G\rangle$$

$$- \sum_{p\in R} t_{pb}^* v_p d_{b\downarrow}^\dagger \gamma_{p0}^\dagger \sum_{k\in L} t_{ka} t_{lb} \frac{u_k v_k^*}{(E_k - \epsilon)(-2\epsilon)} |G;0,0;G\rangle$$

$$= + \sum_{k\in L} \sum_{p\in R} t_{ka} t_{lb} t_{pb}^* \frac{u_k v_k^* v_p}{(E_k - \epsilon)(-2\epsilon)} \gamma_{p0}^\dagger d_{b\downarrow}^\dagger |G;0,0;G\rangle$$

$$\xrightarrow{\frac{1-P}{H_0}} + \sum_{k\in L} \sum_{p\in R} t_{ka} t_{lb} t_{pb}^* \frac{u_k v_k^* v_p}{(E_k - \epsilon)(-2\epsilon)(E_p - \epsilon)} \gamma_{p0}^\dagger d_{b\downarrow}^\dagger |G;0,0;G\rangle$$

$$\xrightarrow{H_T} + \sum_{q \in R} t^*_{qa} u^*_q d^\dagger_{a\uparrow} \gamma_{q0} \sum_{k \in L} \sum_{p \in R} t_{ka} t_{lb} t^*_{pb} \frac{u_k v^*_k v_p}{(E_k - \epsilon)(-2\epsilon)(E_p - \epsilon)} \gamma^\dagger_{p0} d^\dagger_{b\downarrow} |G; 0, 0; G\rangle$$

$$= + \sum_{k \in L} \sum_{p \in R} t_{ka} t_{lb} t^*_{pb} t^*_{pa} \frac{u_k v^*_k u^*_p v_p}{(E_k - \epsilon)(-2\epsilon)(E_p - \epsilon)} d^\dagger_{a\uparrow} d^\dagger_{b\downarrow} |G; 0, 0; G\rangle$$

$$= + t^4 \sum_{k \in L} \sum_{p \in R} \frac{u_k v^*_k u^*_p v_p}{(E_k - \epsilon)(-2\epsilon)(E_p - \epsilon)} |G; \uparrow, \downarrow; G\rangle$$

$$= + \frac{1}{16} J e^{+i\phi} |G; \uparrow, \downarrow; G\rangle \ .$$

The path which transfers Cooper pairs from the left to the right superconductor contributes the complex conjugate of the above one.

Path P_2:

$$|G; \uparrow, \downarrow; G\rangle = d^\dagger_{a\uparrow} d^\dagger_{b\downarrow} |G; 0, 0; G\rangle$$

$$= d^\dagger_{a\uparrow} d^\dagger_{b\downarrow} \prod_k \left(u_k + v_k c^\dagger_{k\uparrow} c^\dagger_{-k\downarrow} \right) \prod_q \left(u_q + v_q c^\dagger_{q\uparrow} c^\dagger_{-q\downarrow} \right) |0; 0, 0; 0\rangle$$

$$\xrightarrow{H_T} - t \sum_{k \in L} u_k \gamma^\dagger_{k0} d_{a\uparrow} d^\dagger_{a\uparrow} d^\dagger_{b\downarrow} |G; 0, 0; G\rangle = - t \sum_{k \in L} u_k \gamma^\dagger_{k0} d^\dagger_{b\downarrow} |G; 0, 0; G\rangle$$

$$\xrightarrow[H_0]{1-P} - t \sum_{k \in L} \frac{u_k}{E_k - \epsilon} \gamma^\dagger_{k0} d^\dagger_{b\downarrow} |G; 0, 0; G\rangle$$

$$\xrightarrow{H_T} - t \sum_{l \in L} u_l \gamma^\dagger_{l1} d_{b\downarrow} (-t) \sum_{k \in L} \frac{u_k}{E_k - \epsilon} \gamma^\dagger_{k0} d^\dagger_{b\downarrow} |G; 0, 0; G\rangle$$

$$+ t \sum_{l \in L} v^*_l \gamma_{l0} d_{b\downarrow} (-t) \sum_{k \in L} \frac{u_k}{E_k - \epsilon} \gamma^\dagger_{k0} d^\dagger_{b\downarrow} |G; 0, 0; G\rangle$$

$$= + t^2 \sum_{kl \in L} \frac{u_k u_l}{E_k - \epsilon} \gamma^\dagger_{k0} \gamma^\dagger_{l1} |G; 0, 0; G\rangle$$

$$+ t^2 \sum_{k \in L} \frac{u_k v^*_k}{E_k - \epsilon} |G; 0, 0; G\rangle$$

$$\xrightarrow[H_0]{1-P} + t^2 \sum_{kl \in L} \frac{u_k u_l}{(E_k - \epsilon)(E_k + E_l - 2\epsilon)} \gamma^\dagger_{k0} \gamma^\dagger_{l1} |G; 0, 0; G\rangle$$

$$+ t^2 \sum_{k \in L} \frac{u_k v^*_k}{(E_k - \epsilon)(-2\epsilon)} |G; 0, 0; G\rangle \ .$$

The first term will be projected out at the end.

$$\xrightarrow{H_T} -t \sum_{p \in R} v_p d_{a\uparrow}^\dagger \gamma_{p1}^\dagger \, (+t^2) \sum_{k \in L} \frac{u_k v_k^*}{(E_k - \epsilon)(-2\epsilon)} \, |G; 0,0; G\rangle$$

$$= +t^3 \sum_{k \in L} \sum_{p \in R} \frac{u_k v_k^* v_p}{(E_k - \epsilon)(-2\epsilon)} \, \gamma_{p1}^\dagger d_{a\uparrow}^\dagger \, |G; 0,0; G\rangle$$

$$\xrightarrow{\frac{1-P}{H_0}} +t^3 \sum_{k \in L} \sum_{p \in R} \frac{u_k v_k^* v_p}{(E_k - \epsilon)(-2\epsilon)(E_p - \epsilon)} \, \gamma_{p1}^\dagger d_{a\uparrow}^\dagger \, |G; 0,0; G\rangle$$

$$\xrightarrow{H_T} -t \sum_{q \in R} u_q^* d_{b\downarrow}^\dagger \gamma_{q1} \, (+t^3) \sum_{k \in L} \sum_{p \in R} \frac{u_k v_k^* v_p}{(E_k - \epsilon)(-2\epsilon)(E_p - \epsilon)} \, \gamma_{p1}^\dagger d_{a\uparrow}^\dagger \, |G; 0,0; G\rangle$$

$$= +t^4 \sum_{k \in L} \sum_{p \in R} \frac{u_k v_k^* u_p^* v_p}{(E_k - \epsilon)(-2\epsilon)(E_p - \epsilon)} \, d_{a\uparrow}^\dagger d_{b\downarrow}^\dagger \, |G; 0,0; G\rangle$$

$$= +t^4 \sum_{k \in L} \sum_{p \in R} \frac{u_k v_k^* u_p^* v_p}{(E_k - \epsilon)(-2\epsilon)(E_p - \epsilon)} \, |G; \uparrow, \downarrow; G\rangle$$

$$= +\frac{1}{16} J e^{+i\phi} \, |G; \uparrow, \downarrow; G\rangle \; .$$

The path which transfers Cooper pairs from the left to the right superconductor contributes the complex conjugate of the above one.

Path P_5:

$$|G; \uparrow, \downarrow; G\rangle = d_{a\uparrow}^\dagger d_{b\downarrow}^\dagger \, |G; 0,0; G\rangle$$

$$= d_{a\uparrow}^\dagger d_{b\downarrow}^\dagger \prod_k \left(u_k + v_k c_{k\uparrow}^\dagger c_{-k\downarrow}^\dagger \right) \prod_q \left(u_q + v_q c_{q\uparrow}^\dagger c_{-q\downarrow}^\dagger \right) |0; 0,0; 0\rangle$$

$$\xrightarrow{H_T} -t \sum_{k \in L} u_k \gamma_{k0}^\dagger d_{a\uparrow} \, d_{a\uparrow}^\dagger d_{b\downarrow}^\dagger \, |G; 0,0; G\rangle = -t \sum_{k \in L} u_k \gamma_{k0}^\dagger d_{b\downarrow}^\dagger \, |G; 0,0; G\rangle$$

$$\xrightarrow{\frac{1-P}{H_0}} -t \sum_{k \in L} \frac{u_k}{E_k - \epsilon} \, \gamma_{k0}^\dagger d_{b\downarrow}^\dagger \, |G; 0,0; G\rangle$$

$$\xrightarrow{H_T} -t \sum_{l \in L} u_l \gamma_{l1}^\dagger d_{b\downarrow} \, (-t) \sum_{k \in L} \frac{u_k}{E_k - \epsilon} \, \gamma_{k0}^\dagger d_{b\downarrow}^\dagger \, |G; 0,0; G\rangle$$

$$+ t \sum_{l \in L} v_l^* \gamma_{l0} d_{b\downarrow} \, (-t) \sum_{k \in L} \frac{u_k}{E_k - \epsilon} \, \gamma_{k0}^\dagger d_{b\downarrow}^\dagger \, |G; 0,0; G\rangle$$

$$= +t^2 \sum_{kl \in L} \frac{u_k u_l}{E_k - \epsilon} \, \gamma_{k0}^\dagger \gamma_{l1}^\dagger \, |G; 0,0; G\rangle$$

$$+ t^2 \sum_{k \in L} \frac{u_k v_k^*}{E_k - \epsilon} \, |G; 0,0; G\rangle$$

$$\xrightarrow{\frac{1-P}{H_0}} + t^2 \sum_{kl\in L} \frac{u_k u_l}{(E_k - \epsilon)(E_k + E_l - 2\epsilon)} \gamma_{k0}^\dagger \gamma_{l1}^\dagger |G; 0, 0; G\rangle$$

$$+ t^2 \sum_{k\in L} \frac{u_k v_k^*}{(E_k - \epsilon)(-2\epsilon)} |G; 0, 0; G\rangle \; .$$

The first term will be projected out at the end.

$$\xrightarrow{H_T} - t \sum_{p\in R} u_p^* d_{b\uparrow}^\dagger \gamma_{p0} \, (+t^2) \sum_{k\in L} \frac{u_k v_k^*}{(E_k - \epsilon)(-2\epsilon)} |G; 0, 0; G\rangle$$

$$- t \sum_{p\in R} v_p d_{b\uparrow}^\dagger \gamma_{p1}^\dagger \, (+t^2) \sum_{k\in L} \frac{u_k v_k^*}{(E_k - \epsilon)(-2\epsilon)} |G; 0, 0; G\rangle$$

$$= + t^3 \sum_{k\in L} \sum_{p\in R} \frac{u_k v_k^* v_p}{(E_k - \epsilon)(-2\epsilon)} \gamma_{p1}^\dagger d_{b\uparrow}^\dagger |G; 0, 0; G\rangle$$

$$\xrightarrow{\frac{1-P}{H_0}} + t^3 \sum_{k\in L} \sum_{p\in R} \frac{u_k v_k^* v_p}{(E_k - \epsilon)(-2\epsilon)(E_p - \epsilon)} \gamma_{p1}^\dagger d_{b\uparrow}^\dagger |G; 0, 0; G\rangle$$

$$\xrightarrow{H_T} - t \sum_{q\in R} u_q^* d_{a\downarrow}^\dagger \gamma_{q1} \, (+t^3) \sum_{k\in L} \sum_{p\in R} \frac{u_k v_k^* v_p}{(E_k - \epsilon)(-2\epsilon)(E_p - \epsilon)} \gamma_{p1}^\dagger d_{b\uparrow}^\dagger |G; 0, 0; G\rangle$$

$$= - t^4 \sum_{k\in L} \sum_{p\in R} \frac{u_k v_k^* u_q^* v_p}{(E_k - \epsilon)(-2\epsilon)(E_p - \epsilon)} d_{a\downarrow}^\dagger d_{b\uparrow}^\dagger |G; 0, 0; G\rangle$$

$$= - t^4 \sum_{k\in L} \sum_{p\in R} \frac{u_k v_k^* u_q^* v_p}{(E_k - \epsilon)(-2\epsilon)(E_p - \epsilon)} |G; \downarrow, \uparrow; G\rangle$$

$$= - \frac{1}{16} J e^{+i\phi} |G; \downarrow, \uparrow; G\rangle \; .$$

The path which transfers Cooper pairs from the left to the right superconductor contributes the complex conjugate of the above one.

Adding the contributions of the different paths, we obtain

$$H_{C_1}^{(4)} = 4 \times \frac{J}{16} \times 2\cos\phi \times \begin{bmatrix} 0 & & \\ & -1 & +1 \\ & +1 & -1 \\ & & 0 \end{bmatrix} = J\cos\phi \left[\mathbf{S}_a \cdot \mathbf{S}_b - \frac{1}{4} \right] \; . \tag{46}$$

References

1. See, e.g., *Mesoscopic Electron Transport*, edited by L. L. Sohn, L. P. Kouwenhoven, and G. Schön (Kluwer, Dordrecht, 1997).
2. I. O. Kulik, Sov. Phys. JETP **22**, 841 (1966); H. Shiba and T. Soda, Prog. Theor. Phys. **41**, 25 (1969); L. N. Bulaevskii, V. V. Kuzii, and A. A. Sobyanin, JETP Lett. **25**, 290 (1977).

3. L. I. Glazman and K. A. Matveev, Pis'ma Zh. Eksp. Teor. Fiz. **49**, 570 (1989) [JETP Lett. **49**, 659 (1989)]; B. I. Spivak and S. A. Kivelson, Phys. Rev. B **43**, 3740 (1991).
4. D. C. Ralph, C. T. Black, and M. Tinkham, Phys. Rev. Lett. **74**, 3241 (1995).
5. C. B. Whan and T. P. Orlando, Phys. Rev. B **54**, R5255 (1996); A. Levy Yeyati, J. C. Cuevas, A. Lopez-Davalos, and A. Martin-Rodero, Phys. Rev. B **55**, R6137 (1997).
6. S. Ishizaka, J. Sone, and T. Ando, Phys. Rev. B **52**, 8358 (1995); A. V. Rozhkov and D. P. Arovas, Phys. Rev. Lett. **82**, 2788 (1999); A. A. Clerk and V. Ambegaokar, Phys. Rev. B **61**, 9109 (2000).
7. K. A. Matveev *et al.*, Phys. Rev. Lett. **70**, 2940 (1993).
8. R. Bauernschmitt, J. Siewert, A. Odintsov, and Yu. V. Nazarov, Phys. Rev. B **49**, 4076 (1994).
9. R. Fazio, F. W. J. Hekking, and A. A. Odintsov, Phys. Rev. B **53**, 6653 (1996).
10. J. Siewert and G. Schön, Phys. Rev. B **54**, 7424 (1996).
11. D. Loss and D. P. DiVincenzo, Phys. Rev. A **57**, 120 (1998).
12. D. Loss and E. V. Sukhorukov, Phys. Rev. Lett. **84**, 1035 (2000).
13. M.-S. Choi, C. Bruder, and D. Loss, Phys. Rev. B **62**, 13569 (2000).
14. P. Recher, E. V. Sukhorukov, and D. Loss, cond-mat/0009452.
15. A. Auerbach, *Interacting Electrons and Quantum Magnetism* (Springer-Verlag, Berlin, 1994).
16. See, e.g., M. Tinkham, *Introduction to Superconductivity*, 2nd ed. (McGraw-Hill, New York, 1996).
17. D. Vollhardt and P. Wölfle, *The Superfluid Phases of ^3He* (Taylor and Francis, London, 1990).
18. G. M. Luke *et al.*, Nature **394**, 558 (1998).
19. J. M. Kikkawa and D. D. Awschalom, Phys. Rev. Lett. **80**, 4313 (1998).
20. B. J. van Wees and H. Takayanagi, in [1]; H. Takayanagi, E. Toyoda, and T. Akazaki, Superlattices and Microstructures **25**, 993 (1999); A. Chrestin *et al.*, ibid., 711 (1999).

Part II

Quantum Wires

Correlations in Electronic Properties of Semiconductor Quantum Wires

Maura Sassetti[1] and Bernhard Kramer[2]

[1] Dipartamento di Fisica, INFM, Università di Genova, Via Dodecaneso 33
   I-16146 Genova.
[2] I. Institut für Theoretische Physik, Universität Hamburg, Jungiusstraße 9
   D-20355 Hamburg.

Abstract. It is shown that the Raman cross section of semiconductor quantum wires can be completely understood in terms of collective intra- and inter-subband charge and spin modes that propagate with different group velocities. The intensities of the peaks in the cross section are governed by non-analytic, non-Fermi liquid power laws when approaching resonance, with exponents given by the strength of the electron-electron interaction. In addition, results for electron transport through a quasi-one dimensional quantum dot are presented which show that correlations cannot be neglected. However, their signatures depend on which experimental quantity is considered: in the case of linear transport the average of the interaction is measured including the quantum wires to which the quantum dot is connected. In non-linear transport spectroscopy, the interaction is probed at the position of the quantum dot. It is also concluded that in quasi-one dimensional quantum dots the excitations with the lowest energies are due to spin modes.

1   Introduction

In one-dimensional (1D) electron systems electron-electron interaction can be treated exactly by using the bosonization technique in the low-energy limit within the Tomonaga-Luttinger liquid (TLL) model [1–4]. The lowest-energy excitations are collective. The only existing pair excitations are charge (CDE) and spin (SDE) density modes.

The Hamiltonian of the system can be written as a quadratic form,

$$H_0 = \frac{\hbar v_{\mathrm{F}}}{2} \int \mathrm{d}x \left[\Pi_\rho^2(x) + (\partial_x \vartheta_\rho(x))^2 \right]$$

$$+ \frac{1}{\pi} \int \mathrm{d}x \int \mathrm{d}x' \, \partial_x \vartheta_\rho(x) \, V(x - x') \, \partial_{x'} \vartheta_\rho(x')$$

$$+ \frac{\hbar v_\sigma g_\sigma}{2} \int \mathrm{d}x \left[\Pi_\sigma^2(x) + \frac{1}{g_\sigma^2} \left(\partial_x \vartheta_\sigma(x) \right)^2 \right] . \tag{1}$$

Here, the electrons are described by conjugate bosonic fields Π_ρ, ϑ_ρ, and $\Pi_\sigma, \vartheta_\sigma$ which are associated with the collective charge and spin density excitations, respectively. The system length is L and the Fermi velocity v_{F}. The Coulomb interaction, projected to the x-direction, $V(x - x')$, has the Fourier transform

$V(q)$ which appears to be the dominant quantity in the dispersion of the CDE. Their velocity is given by

$$v_\rho = v_F \left(1 - \eta_{ex}^2 + (1 + \eta_{ex})\frac{2V(q)}{\pi \hbar v_F}\right)^{1/2} \equiv \frac{v_F}{g_\rho(q)}. \tag{2}$$

The velocity of the spin modes is determined only by the exchange interaction matrix element $\eta_{ex} \equiv V(2k_F)/2\pi \hbar v_F$, which is usually very small

$$v_\sigma \equiv v_F(1 - \eta_{ex}^2)^{1/2}, \tag{3}$$

The interaction constant for the spin modes is

$$g_\sigma = \left(\frac{1 + \eta_{ex}}{1 - \eta_{ex}}\right)^{1/2}. \tag{4}$$

While the frequency-wavenumber dispersion relations of the CDE are renormalized by $V(q)$, and by the exchange interaction, the SDE energy are influenced only by the exchange. Since this latter interaction is very small the SDE have approximately the dispersion of the pair excitations of the free electrons,

$$\omega_\rho(q) = v_\rho |q|,$$

$$\omega_\sigma(q) = v_\sigma |q| \approx v_F|q|. \tag{5}$$

The quantity $g \equiv g_\rho(q = 0) < 1$ measures the strength of the electron-electron repulsion. Without interaction, $g = 1$; for *Coulomb* repulsion, there is a logarithmic singularity for $q \to 0$ such that $g \to 0$ [4].

Also correlation functions as, for instance, the one-electron Green function can be determined. Generally, they are found to behave according to power laws as a function of some variable ε, say, $C(\varepsilon) \propto \varepsilon^{\mu(g)}$, with a non-integer exponent μ that depends on the interaction parameter g. This is denoted as TLL-behaviour. Wellknown examples are photoemission spectra and one-photon absorption [5].

Similar to the Fermi liquid (FL), the TLL is of fundamental importance in modern theory of condensed matter. Thus, directly measuring the non-analytic behaviour is very important. However, unequivocal experimental evidence is still missing, though several recent results indicate that TTL behaviour is of importance in quantum wires.

On the one hand, Raman scattering data indicate the importance of electron interaction via the observed dispersion of the plasmon modes. They reveal spectral features that are consistent with the above CDE and SDE obtained within the TLL model. However, additional peaks have been observed that seem at first glance not to match theoretical predictions. These are the "single particle excitations" ("SPE"). They appear in perpendicular as well as parallel polarization of incident and scattered light, roughly at the frequencies of the SDE [6–8]. So far, experimentally determined cross sections have not yet been analysed in terms of non-Fermi liquid behaviour.

On the other hand, in spite of many efforts during the last decade that included experiments performed on quasi-1D conductors and superconductors, DC-conductance of quantum wires and fractional quantum Hall states evidence for TLL-behaviour has been only partially convincing [9–14]. The predicted correlation-induced renormalization of the quantum conductance in clean quantum wires has not been found, due to the influence of the leads and the contacts.

Indications of the presence of the correlations have been reported, however, in dirty quantum wires. In the conducting regime, a temperature dependent renormalization of the conductance steps has been found [10] which has been associated with the scattering at impurities [15]. In carbon nanotubes, a temperature dependence of the conductance has been detected [11] which was assigned to the typical non-analytical power-law behaviour predicted earlier for transport through an impurity in a TLL [16].

In particular, transport experiments on quantum wires with the density of the electrons depleted to such a degree that accidentally a quasi-1D quantum dot has been formed [12], evidence has been reported that the temperature dependence of the strength of the Coulomb blockade peaks in the linear conductance show TLL-behaviour

$$\Gamma(T) \propto T^\lambda, \tag{6}$$

with $\lambda > 0$ non-integer, instead of being temperature independent as predicted by the Fermi liquid model. However, the quantitative aspects of these results have been found far from being satisfactory: the interaction constant g predicted from the charging energy has been found different from that determined from the temperature dependence of the peaks [17]. In addition, the number of the low-energy excited many-body states in the quantum dot was found much larger than that expected from the interaction constant.

In this contribution, we summarize the theoretical evidences that indicate the importance of TTL-correlations in the electronic properties of quasi-1D semiconductor quantum wires. We stress that the Raman scattering cross section of a quantum wire can be completely understood in terms of the CDE and SDE of the TLL model, including the peaks that have formerly been denoted as "single particle excitations". In addition, we show that the intensities of the peaks in the cross section show TLL power-law bahaviours.

Furthermore, we report results of the theory of the Coulomb blockade peaks of a quasi-1D quantum dot connected to TLL including the influence of the spin excitations. This will be seen to improve on the consistency between theory and experiment, but still cannot explain the linear and the non-linear transport spectra consistently. We will argue on the basis of these and earlier results that spin excitations are of crucial importance for the understanding of the spectra.

2   Collective Modes in the Raman Cross Section

In general, state-of-the-art semiconductor quantum wires support several electronic subbands. As a consequence, intra- as well as inter-subband CDE and

SDE do exist. They have been investigated experimentally in great detail by using inelastic (Raman) scattering of light.

2.1 Intra-subband Excitations

The intra-subband modes, for which $\omega_\nu(q \to 0) \to 0$, $(\nu = \rho, \sigma)$ can be described by the TLL senario by using the bosonization method, when neglecting the higher subbands [18]. It is found that *all* observed spectral features can be described within this model, consistent with experiments [19].

In particular, in non-resonant scattering, when the energy of the photons, $\hbar\omega_i$, is small compared with the energy gap, E_G, of the host semiconductor, CDE and SDE are found in parallel and perpendicular polarizations of incident and scattered light, respectively.

When approaching resonance, this "classical selection rule" is no longer valid due to the wavenumber dependence of the optical transition matrix elements [19]. This causes higher-order, "dressed" SDE to appear as sharp peaks in the "wrong", namely the parallel polarization. These are the "SPE".

On the other hand, "dressed" CDE cannot appear in perpendicular polarization as sharp peaks, but only as a broad background. On the one hand, this is due to the fact that excitations observed in perpendicular polarization should always be accompanied by spin flips. On the other hand, the group velocities of SDE and CDE are rather different such that they can propagate coherently together only during very short periods of time. This implies that the corresponding features in frequency space must be broad.

2.2 Inter-subband Excitations

When higher subbands are considered additional approximations are necessary [20]. For instance, the Fermi velocities in the subbands should be approximately equal. In addition, in order to apply the bosonization technique, interaction-induced coupling between intra- and inter-subband modes have to be small.

For two subbands occupied, in addition to the above two intra-subband modes, six additional modes appear. Two of them are again intra-subband CDE and SDE, but consisting of non-symmetric combinations of left- and right-moving densities. Thus, they are almost not influenced by the interaction.

The other four are inter-subband collective excitations. There are two symmetric and non-symmetric inter-subband CDEs with positive and negative group velocities, respectively, and correspondingly two inter-subband SDEs. The symmetric CDE is shifted to higher energy due to the Coulomb repulsion (depolarization shift). Due to the exchange interaction, the two SDEs are non-degenerate at $q = 0$ [21]. Even far from resonance, the intensities of the corresponding peaks in the Raman cross section depend strongly on wavenumber and energy [22].

Near resonance, $\hbar\omega_i \approx E_G$, also the inter-subband SDEs can appear as "SPE" in both polarizations. However, only the SDE with *positive* group velocity can couple coherently to other SDE for sufficiently long times, in order to yield a sharp peak in the cross section. The inter-subband SDE with negative group velocity cannot pop up as "SPE" in parallel polarization [23].

2.3 TLL-Behaviour of Intensities of Raman Peaks

Non-Fermi liquid behaviour can be expected for the peaks corresponding to the intra-subband excitations.

Due to the bosonized form of the Hamiltonian, the Raman cross section, $\mathrm{Im}\chi$, can be evaluated explicitly. For the peak that originates in the SDE, in the case of a single occupied band,

$$\mathrm{Im}\chi \approx \delta(\omega - \omega_\sigma)\left[(e_\mathrm{i} \cdot e_\mathrm{o})^2 \mathcal{I}_1 + |e_\mathrm{i} \times e_\mathrm{o}|^2 \mathcal{I}_2\right], \tag{7}$$

and for the peak due to the CDE

$$\mathrm{Im}\chi \approx \delta(\omega - \omega_\rho)(e_\mathrm{i} \cdot e_\mathrm{o})^2 \mathcal{I}_0, \tag{8}$$

with $e_\mathrm{i,o}$ the polarizations of incident and scattered light, respectively. For parallel polarization, only the SDE contributes with a peak in the "wrong" polarization. Closed expressions for the intensities $\mathcal{I}_n$ ($n = 0, 1, 2$) can be derived. They contain the interaction parameter g in the exponents of non-analytic functions and have to be evaluated numerically.

In order to identify TLL behaviour, we consider the peak due to SDE. Similar results can be obtained also for the CDE peak. We consider the regime where $q_\mathrm{int} \ll q_\beta > q$. Here, q_int is the inverse of the range of the interaction, $q_\beta \equiv k_\mathrm{B}T/\hbar v_\mathrm{F}$ the wavenumber that corresponds to the temperature and q the wave number of the excitation. In addition, we assume for convenience an interaction with $g > g_0$ ($g_0 \approx 0.2$). One obtains ($n = 1, 2$)

$$\mathcal{I}_n \propto \left(\frac{q_\mathrm{int}}{Q}\right)^{4(1/n - \mu(g))} \tag{9}$$

for $Q \equiv (E_\mathrm{G} + E_\mathrm{F} - \hbar\omega_\mathrm{i} + \hbar v_\mathrm{F}q/2)/\hbar v_\mathrm{F} > q_\beta$, and with $\mu(g) = (g + 1/g - 2)/8$. For $Q < q_\beta$ one obtains

$$\mathcal{I}_n \propto \left(\frac{q_\mathrm{int}\hbar v_\mathrm{F}}{k_\mathrm{B}T}\right)^{4(1/n - \mu(g))} \tag{10}$$

The ratio $\mathcal{I}_1/\mathcal{I}_2$ is independent of the interaction and $\propto \beta^2 = (1/k_\mathrm{B}T)^2$.

The above implies that by determining the dependence of the peak strengths of SDE in parallel and perpendicular polarizations as functions of the temperature and/or the photon energy close to resonance should provide another, experimentally yet un-explored tool for detecting TLL behaviour in quantum wires.

Traditionally, Raman scattering of interacting electrons has been analysed by using random phase approximation (RPA). This works well in the non-resonant regime, since it gives the same results for the dispersion of the CDE as the TLL model. However, the above non-analytic power law behaviours that appear near resonance cannot be described within the RPA.

3   TLL-Behaviour of Transport Through a Double Barrier

As explained above, Raman scattering strongly indicates that electron-electron interaction is important to understand electronic excitations in quantum wires. Why are they not equally strongly visible in the transport properties? For clean wires, this seems to be due to the presence of screening effects and the influence of contacts and leads. However, when the electron density is depleted to such a degree that the fluctuations of the impurity potential are not negligible compared with the Fermi level signatures of the non-analytic TLL-behaviour have been observed [12]. We will discuss in this section, how the experimental data can be analysed [17].

As a model of a quantum dot connected to two single-band quantum wires we consider two δ-function potential barriers in a TLL, $V_i \delta(x - x_i)$ $(i = 1, 2)$. For evaluating the current-voltage characteristic, one can use the path integral formalism [25]. First one performs a thermal average over the "bulk modes" at $x \neq x_1, x_2$. This gives an effective action which ia a quadratic form in four variables: the fluctuations of the particle and spin numbers within the "quantum dot" formed between the two δ-barriers within the interval $a = x_2 - x_1$, N_ρ^- and N_σ^-, and the numbers of imbalanced particles and spins between the left and right TLL-leads, N_ρ^+ and N_σ^+. The current-voltage characteristic can be evaluated by considering the stationary limit of the charge transfer through the the the quantum dot,

$$I = \frac{e}{2} \lim_{t \to \infty} \langle \dot{N}_\rho^+(t) \rangle . \tag{11}$$

The brackets $\langle \ldots \rangle$ include a thermal average over the collective excitations at $x \neq x_1, x_2$ as well as a statistical average performed with the reduced density matrix for the degrees of freedom at $x = x_1, x_2$.

3.1   Charge and Spin Addition Energies

In the continuum limit $(L \to \infty)$, with the inverse temperature $\beta = 1/k_B T$ one obtains for the effective action

$$S_{\text{eff}}[N_\rho^\pm, N_\sigma^\pm] = \int_0^{\hbar\beta} d\tau \, H_B[N_\rho^\pm, N_\sigma^\pm] +$$

$$+ \sum_{r=\pm} \sum_{\nu=\rho,\sigma} \left[\int_0^{\hbar\beta} \int_0^{\hbar\beta} d\tau d\tau' \, N_\nu^r(\tau) K_\nu^r(\tau - \tau') N_\nu^r(\tau') \right.$$

$$\left. - \delta_{\rho,\nu} \int_0^{\hbar\beta} d\tau \, N_\rho^r(\tau) \mathcal{L}^r(\tau) \right] . \tag{12}$$

Here H_B represents the impurity Hamiltonian (see eq. (23)). The Fourier transforms, at Matsubara frequencies $\omega_n = 2\pi n/\hbar\beta$, of the dissipative kernels $K_\nu^\pm(\tau)$

and of the effective "forces" $\mathcal{L}^\pm(\tau)$ are determined by the dispersion relations of the collective modes Eqs. (5),

$$[K_\nu^\pm(\omega_n)]^{-1} = \frac{8v_\nu g_\nu}{\hbar\pi^2} \int_0^\infty dq \frac{1 \pm \cos[q(x_1 - x_2)]}{\omega_n^2 + \omega_\nu^2(q)}, \tag{13}$$

$$\mathcal{L}^\pm(\omega_n) = \frac{4ev_F}{\hbar\pi^2} K_\rho^\pm(\omega_n) \int_{-\infty}^\infty dx \, \mathcal{E}(x, \omega_n)$$

$$\times \int_0^\infty dq \frac{\cos[q(x - x_2)] \pm \cos[q(x - x_1)]}{\omega_n^2 + \omega_\rho^2(q)}. \tag{14}$$

Both, $K_\nu^\pm$ and $\mathcal{L}^\pm$ contain the collective bulk modes which introduce the interaction effects to be described below. First of all, we note that $K_\nu^+(\omega_n \to 0) = 0$ [25,28]. On the other hand, $K_\nu^-(\omega_n \to 0) \neq 0$. The latter describe the costs in energy for changing the numbers of charges and/or spins on the island between the potential barriers. The corresponding Euclidean action is

$$S_0[N_\rho^-, N_\sigma^-] = \sum_{\nu=\rho,\sigma} \frac{E_\nu}{2} \int_0^{\hbar\beta} d\tau \, (N_\nu^-)^2, \tag{15}$$

with the characteristic energies

$$E_\nu = 2K_\nu^-(\omega_n \to 0); \qquad (\nu = \rho, \sigma). \tag{16}$$

For $\nu = \rho$, this corresponds to the charging energy that has to be supplied/is gained, in order to transfer/remove one charge to/from the island as compared with the mean value, $N_\rho^- = \pm 1$. Correspondingly, for $\nu = \sigma$, the "spin addition energy" E_σ is needed/gained in order to change the spin by exactly $\pm 1/2$. The Coulomb interaction that determines the dispersion relation of the charge excitations, increases considerably the charging energy E_ρ, in comparison with the spin addition energy E_σ, as the latter is only influenced by the (small) exchange interaction. Thus, we always expect $E_\rho > E_\sigma$.

The frequency dependent parts of the kernels describes the dynamical effects of the "external leads" and of the correlated excited states in the quantum dot. Their influence is described by the spectral densities $J_\nu^\pm(\omega)$ that are related via analytic continuation to the imaginary-time kernels $J_\nu^\pm(\omega) = 2K_\nu^\pm(\omega_n \to i\omega)/\pi\hbar$ [26,27]. Due to the non-zero range of the interaction, it is not possible to obtain analytic expressions for these densities. However, one can always extract their limits for $\omega \to 0$,

$$J_\nu^\pm(\omega \to 0) = A_\nu^\pm(g_\nu) \frac{\omega}{4g_\nu}, \tag{17}$$

where $A_\rho^- = g_\rho^4(E_\rho/E_0)^2$ $(E_0 = \hbar\pi v_F/2a)$ and for the three other combinations of indices $A_\nu^\pm = 1$. This limit describes the low-frequency dissipative influence of the continua of charge and spin excitations that exist *only* in the external leads, $x < x_1$ and $x > x_2$. It holds also for finite frequencies which must be, however, smaller than the frequency scale corresponding to the range of the interaction, and smaller then the characteristic excitation energy of the correlated electrons in the dot.

Let us now discuss the driving forces. In general, $\mathcal{L}^{\pm}(\tau)$ depend in a quite complicated way on the dispersion of the collective modes and on the shape of the electric field. In the following, we focus on the DC limit where it is sufficient to evaluate the Fourier components for $\omega_n \to 0$. In this case, the quantity $\mathcal{L}^{+}(\tau)$, which acts on the total transmitted charge, depends only on the integral of the time independent electric field over the entire system, the source-drain voltage $U \equiv \int_{-\infty}^{\infty} dx\, \mathcal{E}(x)$,

$$\mathcal{L}^{+}(\tau) = \frac{eU}{2} . \tag{18}$$

Since $\mathcal{L}^{+}$ is the part of the effective force that generates the current transport, this result generalizes the one obtained previously for only one impurity [25]. It can be easily derived from Eq. (14) by using the relation

$$\frac{e^2 v_{\mathrm{F}}}{\hbar \pi^2} \int_0^{\infty} dq\, \frac{\omega_n (1 \pm \cos qx)}{\omega_n^2 + \omega_\rho^2(q)}$$
$$= \sigma_0(0, \omega_n) \pm \sigma_0(x, \omega_n) . \tag{19}$$

Here, $\sigma_0(x, \omega_n)$ is the frequency dependent non-local conductivity of the Luttinger liquid per spin channel [25], with the DC limit $\sigma_0(x, 0) = g_\rho e^2 / h$.

On the other hand, $\mathcal{L}^{-}(\tau)$ acts on the excess charge on the island, it does *not* generate a current. It depends on the spatial shape of the electric field and can formally be written in terms of the total charge $Q_\mathcal{E}$ accumulated between the points x_1 and x_2 in the absence of the barriers as a consequence of the presence of the DC electric field

$$\mathcal{L}^{-}(\tau) = \frac{E_\rho Q_\mathcal{E}}{e} , \tag{20}$$

where

$$Q_\mathcal{E} = 2 \int_{-\infty}^{\infty} dx'\, \mathcal{E}(x')$$
$$\times \lim_{\omega \to 0} \left[\frac{\sigma_0(x_1 - x', -i\omega) - \sigma_0(x_2 - x', -i\omega)}{i\omega} \right] . \tag{21}$$

Equivalently, this can also be understood in terms of addition energies. By introducing explicitly in Eqs. (14) the dependence on the interval considered when evaluating the addition energies, one easily finds

$$\mathcal{L}^{-}(\tau) = \frac{e}{2} E_\rho (x_1 - x_2)$$
$$\times \int_{-\infty}^{\infty} dx\, \mathcal{E}(x) \left[\frac{1}{E_\rho(x - x_1)} - \frac{1}{E_\rho(x - x_2)} \right] . \tag{22}$$

It is reasonable to assume $x_{2,1} = \pm a/2$. If the effective electric field has inversion symmetry, $\mathcal{L}^-$ vanishes. Without inversion symmetry, the electric field generates an effective charge on the island which will influence the total current via coupling between N_ρ^+ and N_ρ^- due to the impurity Hamiltonian H_B [28]. Physically, this "externally induced charge" may be thought of as being generated by a voltage V_G applied to an external gate which electrostatically influences the charge on the island. Thus, the above Eq. (20) can be interpreted as a term representing the effect of the gate voltage in the phenomelogical theory of Coulomb blockade.

3.2   Transport Properties

In order to calculate the electrical current one has to solve the equations of motion for the variables $N_\nu^\pm$. This has been achieved until now only without taking into account the spin [28]. Here, we discuss the general framework which is sufficient to understand the main results.

For barriers much higher then the charging energy, $V_i \gg E_\rho$, the dynamics is dominated by tunneling events that connect the minima of H_B in the 4D $(N_\rho^+, N_\rho^-, N_\sigma^+, N_\sigma^-)$-space [27]. For equal barriers, $V_1 = V_2 = V$, the impurity Hamiltonian is

$$H_B = V\rho_0 \left[\cos\frac{\pi N_\rho^+}{2} \cos\frac{\pi N_\sigma^+}{2} \cos\frac{\pi(n_0 + N_\rho^-)}{2} \cos\frac{\pi N_\sigma^-}{2} \right.$$

$$\left. + \sin\frac{\pi N_\rho^+}{2} \sin\frac{\pi N_\sigma^+}{2} \sin\frac{\pi(n_0 + N_\rho^-)}{2} \sin\frac{\pi N_\sigma^-}{2} \right]. \tag{23}$$

Here $\rho_0 = 2k_F/\pi$ and $n_0 = \rho_0 a$ are the mean electron density and the mean particle number in the dot, respectively.

The transitions between the minima of this function of four variables correspond to different physical processes of transferring electrons from one side to the other of the quantum dot. At very low temperature, the dominant processes are whose which transfer the electron coherently through the dot. In particular, when the number of particles in the island is an odd integer there will be a spin degeneracy in the dot, $N_\sigma^- = \pm 1$. Then, the island acts as a localized magnetic impurity, similar as in the Kondo effect [16].

On the other hand, if the temperature is higher than the tunneling rate through the single barrier, the dominant processes are sequential tunneling events [28,29]. The transfer of charge occurs via uncorrelated single-electron hops into, and out of the island, associated with a corresponding changes in the total spin. This is precisely the regime that recently has become accessible by using cleaved-edge-overgrowth quantum wires [12]. In this region, the minima which one has to consider correspond to pairs of even and odd integer N_ρ^- and N_σ^-, respectively, and vice versa. The dominant transport processes are those which connect minima via transitions $N_\rho^- \to N_\rho^- \pm 1$ associated with changes of the spin $N_\sigma^- \to N_\sigma^- \pm 1$. For each of these processes also the external charge and the spin change by $N_\nu^+ \to N_\nu^+ \pm 1$.

The degeneracy of these minima is lifted by the charge and spin addition energies E_ρ and E_σ which force the system to select favourable charge and spin states in the island. These selection processes become essential at low temperatures, $k_B T < E_\nu$, when current can flow through the dot only under resonant conditions. The latter can be achieved in experiment by tuning external parameters, like the source drain voltage or a gate voltage, in order to create degenerate charge states in the island.

In linear regime $U \to 0$ for $T = 0$, starting with the island occupied by n electrons, we expect that another electron can enter and leave only if the difference between the ground state energies of $n + 1$ and n electrons is aligned with the chemical potential of the external semi-infinite Luttinger systems. The ground state of an even number of electrons in 1D has the total spin 0. On the other hand, the ground state of an odd number of electrons has the spin $N_\sigma^- = \pm 1$ [30]. This implies the resonance condition

$$\mathcal{U}(n + 1, \pm s_{n+1}) - \mathcal{U}(n, \pm s_n) = 0 \qquad (24)$$

with $\mathcal{U}(n, \pm s_n)$ corresponding to the ground state energies with n particles and total spins $s_n = 0$ (n even), or $s_n = 1/2$ (n odd).

With the above charge, spin and external gate terms (16),(20), these conditions become

$$E_\rho \left(n - n_0 - n_G + \frac{1}{2} \right) + (-1)^n \frac{E_\sigma}{2} = 0 . \qquad (25)$$

The variable $n_G = eV_G \delta / E_\rho$ represents the number of induced particles due to the coupling to a gate at which the voltage V_G is applied, with a proportionality factor δ which can be determined experimentally. In Eq. (24), the zero of energy has been assumed to be the external chemical potential.

From the above expression one can see that the distance of the peaks of the linear conductance when changing the gate voltage are given by $\Delta V_G = (E_\rho + (-1)^n E_\sigma)/e\delta \approx E_\rho/e\delta$ since $E_\rho \gg E_\sigma$. Having independent information on δ one can extract the values of E_ρ and E_σ from experiment.

One can investigate the non-linear current-voltage characteristic by increasing the source drain voltage U. In this case, the current shows the Coulomb staircase associated with transitions between successive ground states of the electrons in the quantum dot. In addition, fine structure appears which reflects the excitation spectra of the correlated electrons. The former, for symmetric voltage drops at the barriers, have a maximum periodicity given by $U = \mu_0(n + 1) - \mu_0(n) = E_\rho + (-1)^{n+1} E_\sigma$ with $\mu_0(n)$ the electrochemical potential of n electrons in the dot.

In the present model, the possible excitations are collective spin and charge modes confined within the island. They have quasi-discrete energy spectra, $\omega_\rho(q_m)$ and $\omega_\sigma(q_m)$, due to the discretization of the wave number $q_m = \pi m/a$ associated with the confinement. The screened Coulomb interaction causes a non-linear dispersion relation for the charge modes in the infinite Luttinger system. This leads to non-equidistant charge excitation energies of the quantum dot,

$$\Delta\epsilon_\rho(q_m) = \hbar\left[\omega_\rho(q_m+1) - \omega_\rho(q_m)\right] . \tag{26}$$

Their explicit values depend on the ratio between the distance a between the barriers and the range of the interaction. For values of a much larger then this range, the first excited states of the plasmon exciations are equidistant and are given by the sound velocity for $q \to 0$, $v_\rho = v_F/g_\rho(0)$,

$$\Delta\epsilon_\rho = \frac{\hbar\pi v_\rho}{a} = \frac{\hbar\pi v_F}{ag_\rho} . \tag{27}$$

In the opposite limit, the non linear dispersion is already affecting strongly the first excitation, resulting in a value smaller than (27).

For the spin excitations, the dispersion in the infinite system is linear and the corresponding quasi-discrete spectrum is equidistant

$$\Delta\epsilon_\sigma = \frac{\hbar\pi v_\sigma}{a} . \tag{28}$$

Until now, we have mainly discussed the characteristic energies associated with the the electron island. In order to evaluate the current, one needs also to take into account the influence of the external leads. This can be done considering the behavior of the spectral densities given in (17). In the sequential tunneling regime, one can demonstrate that only the sum of all these densities determines the transport [28]

$$J(\omega) = \sum_{r=\pm} \sum_{\nu=\rho,\sigma} J_\nu^r(\omega) . \tag{29}$$

The frequency behavior of this eventually detemines the current-voltage characteristics both in the linear and in the non-linear regimes. In particular, it determines the exponent of the power-law dependencies of the current as a function of temperature and/or the bias voltage.

In the interesting region of low temperatures, $k_B T \ll \Delta\epsilon_\nu$, the low-frequency behavior of the spectral density is *only* determined by the continua of the charge and spin excitations in the external leads. From (17) one obtains

$$J(\omega) \approx J_{\text{leads}}(\omega) = \frac{\omega}{2}\left(\frac{1}{g_\sigma} + \frac{1+A_\rho^-}{2g_\rho}\right) . \tag{30}$$

Eq. (30) generalizes the results obtained previously for spinless electrons [29,28], where $J(\omega) \approx \omega/g_\rho$. We can then conclude that the presence of the spin in the leads introduces an effective interaction strength

$$\frac{1}{g_{\text{eff}}} = \frac{1}{4}\left(\frac{1+A_\rho^-}{g_\rho} + \frac{2}{g_\sigma}\right) , \tag{31}$$

which eventually determines the exponent of the power law behaviors. For example, it has been shown for spinless electrons [29,28] that the intrinsic width $\Gamma(T)$

of the linear conductance Coulomb peak depends on the temperature. In the limit $k_B T \ll \Delta \epsilon$ this has been found to be given by the power law $\Gamma(T) \propto T^{1/g_\rho - 1}$. Correspondingly, in the presence of the spin, one finds

$$\Gamma(T) \propto T^{1/g_{\text{eff}} - 1} . \tag{32}$$

3.3   Comparison with Experiment

The above discussion emphasizes the importance of charging and spin energies, and of spin and plasmon excitation energies for the linear and non-linear transport properties. It is therefore important to evaluate these quantities microscopically and determine their dependencies on the model parameters. This has been done by using specific models for the interaction potential [17].

The charge and spin addition energies, E_ρ and E_σ, respectively, can be evaluated from the dispersion relations Eqs. (5). The latter,

$$E_\sigma = \frac{\pi \hbar}{2a} \frac{v_\sigma}{g_\sigma} , \tag{33}$$

is almost the same as the addition energy of non-interacting particles and it is related to the Pauli principle due to the quasi- discretization of the spin energies in the dot.

The charging energy E_ρ as a function of the distance between the potential barriers has been evaluated numerically for different interactions [17].

For small a it diverges. Asymptotically, for large a one obtains for a Coulomb interaction, with dielectric constant ϵ, screened by a metallic plane at distance D from a wire with a diameter d ($E_0 = \pi \hbar v_F / 2a$, $\eta = e^2/(2\pi^2 \epsilon \hbar v_F)$),

$$\frac{E_\rho}{E_0} \approx \left[1 + \eta \gamma + 2\eta \ln \left(\frac{2D}{d} \right) \right] \equiv \frac{1}{g_\rho^2} \quad (D < a) , \tag{34}$$

$$\frac{E_\rho}{E_0} \approx \left[1 - \eta \gamma + 2\eta \ln \left(\frac{2a}{\pi d} \right) \right] \equiv \frac{1}{f_\rho^2} \quad (D > a) . \tag{35}$$

Most importantly, E_ρ is only weakly influenced by D/d: changing D/d by a factor of hundred results only in a change of about 30% in E_ρ.

The charging energy always increases with increasing interaction range D, because the cost of energy for putting additional electrons into the island increases. Interaction ranges much larger than a do not change the charging energy because only the short-range part of the repulsion between the electrons contributes. Therefore, E_ρ approaches an asymptotic value. For strong Coulomb interaction ($\eta \gg 1$) and $a \gg d$, this is (cf. Eq. (35))

$$E_\rho = \frac{e^2}{2\pi \epsilon a} \log \left(\frac{2a}{d} \right) \equiv \frac{e^2}{C} . \tag{36}$$

Here, the C is the classical self-capacitance of a cylinder of the length a. The stronger the interaction (increasing η) the larger is E_ρ, very similar to the behaviour of a classical capacitor with a dielectric, $C \propto \epsilon$.

As mentioned above, non-linear transport shows Coulomb steps with width given approximately by E_ρ, and fine structure due to the excited states. It can be then very useful to determine how many excited states are present within the energy window of $2E_\rho$.

We expect to observe within the windows E_ρ more levels due to spin excitations than due to charge ecitations. The ratio $E_\rho/\Delta\epsilon_\nu$ always increases when g_ρ decreases (increasing D). It saturates for $D \gg a$,

$$\frac{E_\rho}{\Delta\epsilon_\rho} = \frac{1}{2f_\rho} \; ; \qquad \frac{E_\rho}{\Delta\epsilon_\sigma} = \frac{v_F}{v_\sigma} \frac{1}{2f_\rho^2} \; . \tag{37}$$

In the opposite limit, $D \ll a$ ($g_\rho \to 1$), the behavior is described by the asymptotic expressions

$$\frac{E_\rho}{\Delta\epsilon_\rho} = \frac{1}{2g_\rho} \; ; \qquad \frac{E_\rho}{\Delta\epsilon_\sigma} = \frac{v_F}{v_\sigma} \frac{1}{2g_\rho^2} \; . \tag{38}$$

We compare now the above results with the experimental data [12]. Results of the temperature dependence of the width of the conductance peaks in the region of Coulomb blockade on quantum wires fabricated with the cleaved-edge-overgrowth technique have been reported. Data have been found to be consistent with

$$\Delta\Gamma(T) \propto T^{1/g^* - 1}, \tag{39}$$

with $g^* \approx 0.82$ and $g^* \approx 0.74$ for a peak closer to the onset of the conductance and the next lower peak, respectively. Taking into account the presence of the spin, the g^* values have to be identified with the effective interaction g_{eff} in Eq. (7). Since for the spin excitations $g_\sigma \approx 1$, we find $g_\rho \approx 0.69$ and $g_\rho \approx 0.59$ which are about 15% smaller than the values given above.

In addition, information about the excited energy levels of correlated electrons have been obtained by measuring the non-linear current voltage characteristics. The number of excited levels observed for a given electron number has been estimated to be roughly between 5 and 10.

The data have been interpreted previously by assuming that within the quantum wire, a quantum dot has been accidentally formed between two maxima of the random potential of impurities. The value of the interaction constant estimated from the parameters of the system together with the charging energy, $g_\rho \approx 0.4$, hve been found to be clearly inconsistent with the above values determined from the temperature dependence of the conductance peaks.

Using our microscopic results, we confirm this inconsistency. By playing with the parameters, we found it to be impossible to identify a parameter region where *all* of the above findings could be considered as consistent with each other, although taking into account the corrections due to the spin excitations tends to improve the consistency. In addition, the number of the excited states observed in the non-linear transport has been found to much larger than what one would expect if the above interaction constant determined from the temperature behaviour of the Coulomb peaks was correct also for the description of the quasi-discrete excitation spectra of the 1D quantum dot. In order to explain the excitation spectra, a much smaller interaction constant seems to be necessary.

In our opinion, this can only mean that the data obtained from non-linear transport *spectroscopy* have to be interpreted by using a different interaction constant than that obtained from the *temperature dependence* of the transport, in the linear as well as in the non-linear regime.

4   Conclusion

We have discussed possible evidences for TLL-behaviour in electronic properties of quantum wires.

In the first part we have provided evidence that Raman scattering could be a powerful tool for detecting TLL-behaviour. The dispersions of the CDE and the SDE are clearly seen in experiment to display the nature of the electron-electron interaction. They propagate with different group velocities and show spin-charge separation. On the other hand, the predicted power-law behaviour of the strengths of the peaks in the Raman spectra when parameters as the temperature or the photon energy are changed ramain to be experimentally confirmed.

In the second part we have made an attempt to interpret quantitatively a recent transport experiment which includes linear as well as non-linear transport data in the region of Coulomb blockade of a quasi-1D quantum dot immersed in a quantum wire. We find, in agreement with earlier conclusions [12,17], that it is impossible to consistently fit all the experimental data with the same interaction parameter. It seems that different interaction strengths have to be used for explaining the temperature dependence of the transport data, and those obtained by measuring the excitation spectra in the quantum dot.

In addition, we have predicted that the energetically lowest excitations in the quasi-1D quantum island considered in the experiment are spin excitations, consistent with an early suggestion based on a semi-phenomenological model in which exact diagonalization and rate equations have been combined [30].

This work has been supported by the Deutsche Forschungsgemeinschaft within SFB 508 of the Universität Hamburg, the MURST via Cofinanziamento 98 and by the EU via TMR (FMRX-CT98-0180 and FMRX-CT96-0042).

References

1. S. Tomonaga, Progr. Theor. Phys. **5**, (1950) 544.
2. J. M. Luttinger, J. Math. Phys. **4**, (1963) 1154.
3. F. D. M. Haldane, J. Phys. C **14**, (1981) 2585.
4. H. J. Schulz, Phys. Rev. Lett. **71**, (1993) 1864.
5. J. Voit, Rep. Progr. Phys. **57**, (1995) 977.
6. A. R. Goñi et al., Phys. Rev. Lett. **67**, (1991) 3298; A. Schmeller et al., Phys. Rev. B **49**, (1994) 14778.
7. C. Schüller et al., Phys. Rev. B **54**, (1996) R17304.

8. F. Perez, B. Jusserand, B. Etienne, Physica E **7**, (2000) 521.
9. T. Ogawa, Physica B **249-251**, (1998) 185.
10. S. Tarucha, T. Honda, T Saku, Sol. St. Comm. **94**, (1995) 413.
11. M. Bockrath et al., Nature **397**, (1999) 598.
12. O. M. Auslaender et al., Phys. Rev. Lett. **84**, (2000) 1764.
13. F. P. Millikan, C. P. Umbach, R. A. Webb, Sol. St. Comm. **97**, (1996) 309.
14. A. M. Chang, L. N. Pfeiffer, K. W. West, Phys. Rev. Lett. **77**, (1996) 2538.
15. D. L. Maslov, Phys. Rev. B **52**, (1995) R14368.
16. C. L. Kane and M. P. A. Fisher, Phys. Rev. B **46**, (1992) 15233.
17. T. Kleimann, M. Sassetti, B. Kramer, A. Yacoby, Phys. Rev. B **62**, (2000) 8144.
18. M. Sassetti, B. Kramer, Adv. Phys. (Leipzig) **7**, (1998) 508.
19. M. Sassetti, B. Kramer, Phys. Rev. Lett. **80**, (1998) 1485.
20. M. Sassetti, B. Kramer, Eur. Phys. J. B **4**, (1998) 357.
21. M. Sassetti, F. Napoli, B. Kramer, Phys. Rev. B **59**, (1999) 7297.
22. M. Sassetti, F. Napoli, B. Kramer, Eur. Phys. J. B **11**, (1999) 643.
23. E. Mariani, M. Sassetti, B. Kramer, Europhys. Lett. **49**, (2000) 224; Ann. Phys. (Leipzig) **8**, (1999) 161.
24. B. Kramer, M. Sassetti, Phys. Rev. B **62**, (2000) 4238.
25. M. Sassetti, B. Kramer, Phys. Rev. B**54**, (1996) R5203.
26. A. Furusaki, N. Nagaosa, Phys. Rev. B**47**, (1993) 3827.
27. M. Sassetti, F. Napoli, U. Weiss, Phys. Rev. B**52**, (1995) 11213.
28. A. Braggio, M. Grifoni, M. Sassetti, Europhys. Lett. **50**, (2000) 136.
29. A. Furusaki, Phys. Rev. B **57**, (1998) 3827.
30. D. Weinmann, W. Häusler, B. Kramer, Phys. Rev. Lett. **74**, (1996) 652.

Fluctuations and Superconductivity
in One Dimension: Quantum Phase Slips

Dmitri S. Golubev and Andrei D. Zaikin

[1] Forschungszentrum Karlsruhe, Institut für Nanotechnologie, 76021 Karlsruhe, Germany
[2] I.E.Tamm Department of Theoretical Physics, P.N.Lebedev Physics Institute, Leninskii pr. 53, 117924 Moscow, Russia

Abstract. We investigate quantum fluctuations of the superconducting order parameter in thin homogeneous superconducting wires at all temperatures below T_C. We have derived the effective rate for quantum phase slips in such wires. This rate determines the resistance of the wire. In very thin and dirty metallic wires the effect is shown to be clearly observable even at lowest temperatures. Our results are fully consistent with recent experimental findings which provide direct evidence for the effect of quantum phase slips.

1 Introduction

It is well established that superconducting fluctuations play a very important role in reduced dimension. Above the critical temperature T_C such fluctuations yield an enhanced conductivity [1]. Below T_C fluctuations are known to destroy the long-range order in low dimensional superconductors [2]. Does the latter result mean that the resistance of such superconductors always remains finite (or even infinite), or can it drop to zero under certain conditions?

It was first pointed out by Little [3] that quasi-one-dimensional wires made of a superconducting material can acquire a finite resistance below T_C of a bulk material due to the mechanism of thermally activated phase slips (TAPS). This TAPS process corresponds to local destruction of superconductivity by thermal fluctuations. Superconducting phase $\varphi(t)$ can flip by 2π across those points of the wire where the order parameter is (temporarily) destroyed. According to the Josephson relation $V = \hbar\dot{\varphi}/2e$ such phase slips cause a nonzero voltage and, hence, dissipative currents inside the wire. A theory of this TAPS phenomenon was provided in Ref. [4]. This theory yields a natural result, that the TAPS probability and, hence, resistance of a superconducting wire R below T_C are determined by the activation exponent

$$R(T) \propto \exp(-U/T), \qquad U \sim \frac{N_0 \Delta^2(T)}{2} s\xi(T), \qquad (1)$$

where $U(T)$ is the effective potential barrier for TAPS determined simply as the superconducting condensation energy (N_0 is the metallic density of states at the Fermi energy and $\Delta(T)$ is the BCS order parameter) for a part of the wire of a volume $s\xi$ where superconductivity is destroyed by thermal fluctuations

(s is the wire cross section and $\xi(T)$ is the superconducting coherence length). At temperatures very close to T_C eq. (1) yields appreciable resistivity which was indeed detected experimentally [5]. Close to T_C the experimental results [5] fully confirm the activation behavior of $R(T)$ predicted in eq. (1). However, as the temperature is lowered further below T_C the number of TAPS decreases exponentially and no measurable resistance is predicted by the theory [4] outside of immediate vicinity of the critical temperature.

Experiments [5] were done on small diameter whiskers and thin film samples of typical diameters ~ 5000 angstrom. Recent progress in nanolithographic technique allowed to fabricate samples with much smaller diameters down to ~ 10 nm. In such systems one can consider a possibility for phase slips to be created not only due to thermal but also due to *quantum* fluctuations of a superconducting order parameter. Mooij and coworkers [6] discussed this possibility and attempted to observe quantum phase slips (QPS) experimentally.

Later Giordano [7] performed experiments which clearly demonstrated a notable resistivity of ultra-thin superconducting wires far below T_C. There observations could not possibly be explained by the TAPS theory [4] and were attributed to QPS. Other groups also reported noticeable deviations from the TAPS prediction in thin (quasi-)1D wires [8,9].

Initially a natural explanation of these observations in terms of QPS faced certain problems: a first estimate [7] for the QPS tunneling rate $\propto \exp(-S_{QPS})$, however, leads to the disappointing conclusion that the action S_{QPS} roughly equals to the number of transverse channels $N_{Ch} = k_F^2 s$ in the wire, which is very large even for the thinnest wires used in the experiments [7] (e.g. for $s \sim 10^{-12}$ cm^2 we have $S_{QPS} \sim 10^2 \div 10^3$), and therefore QPS effects should be strongly suppressed. This estimate is obtained from the formula $S_{QPS} \sim U_{QPS}/\omega_a$, with an attempt frequency $\omega_a \sim \Delta$. Assuming U_{QPS} to be the condensation energy $N_0\Delta^2/2$ in a volume $s\xi_0 \sim sv_F/\Delta$ during a time Δ^{-1}, one obtains $S_{QPS} \sim s\xi_0 N_0\Delta^2/2\Delta \sim k_F^2 s/4\pi^2 \sim N_{Ch}$. A similar estimate has been obtained using a phenomenological time dependent Ginzburg-Landau (TDGL) free energy with second order time derivatives [10,11]. Furthermore, Duan [11] claimed an additional suppression of the QPS rate by a factor $\sim \exp(-100)$ due to the effect of electromagnetic environment. Since dissipation is known to reduce the probability of quantum tunneling [12], dissipative currents could cause even further suppression of the QPS process. In contrast, in order to explain the magnitude of the resistance for the thinnest wires measured in Ref. [7] one needs to have $S_{QPS} \sim 10$ with the QPS rate by orders of magnitude larger than that derived from the above estimates.

In Ref. [13] it was argued that the above estimates need qualitative improvement. First, the estimate for the potential barrier should be trivially improved upon: as the typical electron mean free path in the wires [7] is very small (typically below 10 nm), one should rather take $\xi \sim \sqrt{l\xi_0} \ll \xi_0$ for the typical QPS size. Second, it was demonstrated [13] that the role of the electromagnetic field for thin wires was overestimated in Ref. [11] (roughly by a factor $\sqrt{s}/\lambda_L \sim 10^{-1} \div 10^{-2}$, λ_L is the London length of a bulk superconductor).

Third, the dissipative currents turn out not to have a strong impact on the QPS rate, especially in the limit of low T. Also from a general point of view, TDGL-based theories are known to be insufficient to describe non-stationary processes in superconductors and in many cases fail to give qualitatively correct results. Therefore in Refs. [13,14] a microscopic theory of QPS processes was developed with the aid of the imaginary time effective action technique [15] which accounts for non-equilibrium, dissipative and electromagnetic effects during a QPS event.

One of the main conclusions reached in Refs. [13,14] is that the QPS probability is considerably larger than it was predicted previously. For ultra-thin superconducting wires with sufficiently many impurities and with diameters in the 10 nm range this probability can already be large enough to yield experimentally observable phenomena. Also, further interesting effects including quantum phase transitions caused by interactions between quantum phase slips were discussed [13,14].

In spite of all these developments an unambiguous interpretation of the results [7] in terms of QPS could still be questioned because of possible granularity of the samples used in these experiments. In the case of granular superconducting chains QPS can be easily created inside weak links connecting neighbouring grains. Also in this case superconducting fluctuations might play an important role [16], however – in contrast to the QPS scenario [13,14] – the superconducting order parameter *needs not to be destroyed* during the QPS event.

Recently, Bezryadin, Lau and Tinkham [17] came up with a new technology which allowed them to fabricate essentially *uniform* superconducting wires with thicknesses down to 3–5 nm. According to our theory [13,14] the QPS effects should be sufficiently large in such systems to be observed in experiments. And indeed, the authors [17] observed that several wires showed no sign of superconductivity even at temperatures well below the bulk critical temperature. Moreover, at lower temperatures their resistance was found to actually *increase* with decreasing temperature, i.e. these samples could even turn insulating at $T \to 0$. The authors [17] also argued that their experimental data can be interpreted in terms of a quantum dissipative phase transition [18,19] which was also predicted in [14] to occur in ultra-thin superconducting wires for a certain parameter range.

The results [17] are qualitatively consistent with previous experimental findings [7]. Both experimental works clearly support our general understanding of the role of QPS processes in mesoscopic superconducting wires and call for a more detailed theoretical study of the QPS effects. In Refs. [13,14] an importance of collective modes [20] and QPS interaction effects was mainly emphasized. These are particularly important for long wires. In this paper we will present a detailed microscopic investigation of single quantum phase slips. We will focus our attention on an accurate evaluation of the QPS tunneling rate rather than on the interaction effects between different phase slips discussed in Refs. [13,14]. We will go beyond the exponential accuracy and also estimate a pre-exponential factor for the QPS rate. We will then use our results for a direct quantitative comparison with experimental results [17].

The structure of the paper is as follows. In Section 2 we will formulate a simple derivation of the effective action for our problem with an emphasis put on the Ward identities. In Section 3 we will make use of our general results and derive the action for a special case of ultra-thin superconducting wires. In Section 4 the QPS rate is evaluated within the exponential accuracy while Section 5 is devoted to an estimate of the pre-exponential factor for this rate. Comparison with experiments and brief conclusions are presented in Section 6.

2   The Model and Effective Action

The starting point for our analysis is a model Hamiltonian that includes a short range attractive BCS and a long range repulsive Coulomb interaction. The idea is to integrate out the electronic degrees of freedom on the level of the partition function, so that we are left with an effective theory in terms of collective fields [19,21,22]. The partition function Z is conveniently expressed as a path integral over the anticommuting electronic fields $\bar{\psi}$, ψ and the commuting gauge fields V and $\mathbf{A}$, with Euclidean action

$$
S = \int dx \Big(\bar{\psi}_\sigma [\partial_\tau - ieV + \xi(\nabla - \frac{ie}{c}\mathbf{A})]\psi_\sigma -
$$
$$
-\lambda \bar{\psi}_\uparrow \bar{\psi}_\downarrow \psi_\downarrow \psi_\uparrow + ienV + [\mathbf{E}^2 + \mathbf{B}^2]/8\pi \Big) . \tag{2}
$$

Here $\xi(\nabla) \equiv -\nabla^2/2m - \mu + U(x)$ describes a single conduction band with quadratic dispersion and also includes an arbitrary impurity potential, λ is the BCS coupling constant, $\sigma =\uparrow,\downarrow$ is the spin index, and en denotes the background charge density of the ions. In our notation dx denotes $d^3x d\tau$ and we use units in which $\hbar$ and k_B are set equal to unity. The field strengths are functions of the gauge fields through $\mathbf{E} = -\nabla V + (1/c)\partial_\tau \mathbf{A}$ and $\mathbf{B} = \nabla \times \mathbf{A}$ in the usual way for the imaginary time formulation.

We use a Hubbard-Stratonovich transformation to decouple the BCS interaction term and to introduce the superconducting order parameter field $\tilde{\Delta} = \Delta e^{i\varphi}$

$$
\exp\Big(\lambda \int dx \bar{\psi}_\uparrow \bar{\psi}_\downarrow \psi_\downarrow \psi_\uparrow \Big) = \Big[\int \mathcal{D}^2 \tilde{\Delta} e^{-\frac{1}{\lambda}\int dx \Delta^2} \Big]^{-1}
$$
$$
\times \int \mathcal{D}^2 \tilde{\Delta} e^{-\int dx (\frac{1}{\lambda}\Delta^2 + \tilde{\Delta}\bar{\psi}_\uparrow \bar{\psi}_\downarrow + \tilde{\Delta}^* \psi_\downarrow \psi_\uparrow)} , \tag{3}
$$

where the first factor is for normalization and will not be important in the following. As a result, the partition function now reads

$$
Z = \int \mathcal{D}^2 \tilde{\Delta} \int \mathcal{D}^3 \mathbf{A} \int \mathcal{D}V \mathcal{D}^2 \Psi e^{(-S_0 - \int dx \bar{\Psi}\mathcal{G}^{-1}\Psi)} , \tag{4}
$$
$$
S_0[V, \mathbf{A}, \Delta] = \int dx \Big(\frac{\mathbf{E}^2 + \mathbf{B}^2}{8\pi} + ienV + \frac{\Delta^2}{\lambda} \Big) ,
$$

where the Nambu spinor notation for the electronic fields and the matrix Green function in Nambu space

$$\Psi = \begin{pmatrix} \psi_\uparrow \\ \psi_\downarrow \end{pmatrix}, \quad \bar{\Psi} = (\bar{\psi}_\uparrow \; \psi_\downarrow) \; ;$$

$$\tilde{\mathcal{G}}^{-1} = \begin{pmatrix} \partial_\tau - ieV + \xi(\nabla - \frac{ie}{c}\mathbf{A}) & \tilde{\Delta} \\ \tilde{\Delta}^* & \partial_\tau + ieV - \xi(\nabla + \frac{ie}{c}\mathbf{A}) \end{pmatrix}. \tag{5}$$

has been introduced. After the Gaussian integral over the electronic degrees of freedom, we are left with the final effective action

$$S_{\text{eff}} = -\text{Tr}\ln\tilde{\mathcal{G}}^{-1} + S_0[V, \mathbf{A}, \Delta]. \tag{6}$$

Here the trace Tr denotes both a matrix trace in Nambu space and a trace over internal coordinates or momenta and frequencies. In the following "tr" is used to denote a trace over internal coordinates only.

The gauge invariance of the theory enables us to rewrite the action (6) in a different form, which is more convenient for us,

$$S_{\text{eff}} = -\text{Tr}\ln\mathcal{G}^{-1} + S_0[V, \mathbf{A}, \Delta], \tag{7}$$

where

$$\mathcal{G}^{-1} = \begin{pmatrix} \partial_\tau + \xi(\nabla) - ie\Phi + \frac{m\mathbf{v}_s^2}{2} - \frac{i}{2}\{\nabla, \mathbf{v}_s\} & \Delta \\ \Delta & \partial_\tau - \xi(\nabla) + ie\Phi - \frac{m\mathbf{v}_s^2}{2} - \frac{i}{2}\{\nabla, \mathbf{v}_s\} \end{pmatrix}, \tag{8}$$

and we have introduced the gauge invariant linear combinations of the electromagnetic potentials and the phase of the order parameter

$$\Phi = V - \frac{\dot{\varphi}}{2e}, \quad \mathbf{v}_s = \frac{1}{2m}\left(\nabla\varphi - \frac{2e}{c}\mathbf{A}\right). \tag{9}$$

The brackets $\{.,.\}$ denote an anti-commutator.

2.1   Perturbation Theory

The action (7) cannot be evaluated exactly. Here we will perform a perturbative expansion in Φ and $\mathbf{v}_s$. We will keep the terms up to the second order in these values. This perturbation theory is sufficient for nearly all practical purposes, because nonlinear electromagnetic effects (described by higher order terms) are known to be usually very small in the systems in question. Our general derivation holds for an arbitrary concentration and distribution of impurities as well as for arbitrary fluctuations of the order parameter field in space and time.

We split the inverse Green function (8) into two parts

$$\mathcal{G}_0^{-1} = \begin{pmatrix} \partial_\tau + \xi(\nabla) & \Delta \\ \Delta & \partial_\tau - \xi(\nabla) \end{pmatrix}, \tag{10}$$

and

$$\mathcal{G}_1^{-1} = \begin{pmatrix} -ie\Phi + \frac{m\mathbf{v}_s^2}{2} - \frac{i}{2}\{\nabla, \mathbf{v}_s\} & 0 \\ 0 & ie\Phi - \frac{m\mathbf{v}_s^2}{2} - \frac{i}{2}\{\nabla, \mathbf{v}_s\} \end{pmatrix}. \tag{11}$$

The logarithm in the equation (7) can now be expanded in powers of $\mathcal{G}_1^{-1}$ and we get

$$\mathrm{Tr}\ln\mathcal{G}^{-1} = \mathrm{Tr}\ln\mathcal{G}_0^{-1} + \mathrm{Tr}(\mathcal{G}_0\mathcal{G}_1^{-1}) - \frac{1}{2}\mathrm{Tr}(\mathcal{G}_0\mathcal{G}_1^{-1})^2. \tag{12}$$

The Green function $\mathcal{G}_0$ has the form

$$\mathcal{G}_0 = \begin{pmatrix} G & F \\ F & \bar{G} \end{pmatrix}. \tag{13}$$

In eq. (13) we used the fact that the non-diagonal component Δ in the matrix $\mathcal{G}_0^{-1}$ is real. As a result we have $\bar{F} = F$, $F(x_1, x_2) = F(x_2, x_1)$ and $\bar{G}(x_1, x_2) = -G(x_2, x_1)$.

2.2   Ward Identities

The Green function $\mathcal{G}_0$ satisfies an important identity, which is easy to check:

$$\mathcal{G}_0^{-1}\chi - \chi\mathcal{G}_0^{-1} = \frac{\partial\chi}{\partial\tau} - \{\nabla, \frac{\nabla\chi}{2m}\}\sigma_3, \tag{14}$$

where χ is an arbitrary function of time and space, and σ_3 is one of the Pauli matrices. Multiplying this matrix identity by $\mathcal{G}$ from the left and from the right side and taking the diagonal components of the resulting matrix equation we get two identities:

$$\chi G - G\chi = G\left(\dot{\chi} - \{\nabla, \frac{\nabla\chi}{2m}\}\right)G + F\left(\dot{\chi} + \{\nabla, \frac{\nabla\chi}{2m}\}\right)F,$$

$$\chi\bar{G} - \bar{G}\chi = F\left(\dot{\chi} - \{\nabla, \frac{\nabla\chi}{2m}\}\right)F + \bar{G}\left(\dot{\chi} + \{\nabla, \frac{\nabla\chi}{2m}\}\right)\bar{G}. \tag{15}$$

Below we will use these identities in order to decouple the effective action of the BCS superconductor and to reduce it to a transparent and convenient form. It is important to emphasize again that these identities are valid for any impurity distribution and for any time and spatial dependence of the order parameter field. It is also worth mentioning that the Ward identity (14) is *not* the result of the gauge invariance of the theory. It remains valid even for uncharged particles.

The Ward identity related to the gauge invariance of our theory has a different form:

$$\mathcal{G}_0^{-1}\sigma_3\chi - \chi\sigma_3\mathcal{G}_0^{-1} = \frac{\partial\chi}{\partial\tau}\sigma_3 - \{\nabla, \frac{\nabla\chi}{2m}\} - 2i\sigma_2\Delta\chi. \tag{16}$$

We will use this identity to transform the first order correction to the action. It is interesting, that in the absence of superconductivity the identities (14) and (16) are equivalent because the inverse Green function commutes with σ_3 in this case. For superconductors, however, these two identities are different.

2.3 First Order

The first order correction to the effective action is

$$S_1 = -\mathrm{Tr}(\mathcal{G}_0\mathcal{G}_1^{-1}) = -\mathrm{tr}\left[\left(\frac{m\mathbf{v}_s^2}{2} - ie\Phi\right)(G - \bar{G}) - \frac{i}{2}\{\nabla, \mathbf{v}_s\}(G + \bar{G})\right]. \quad (17)$$

With the aid of the Ward identity (16) it easy to show that the phase of the order parameter drops out from the first order terms in the electromagnetic fields. The action S_1 can therefore be rewritten as

$$S_1 = -\mathrm{tr}(m\mathbf{v}_s^2 G) - \int dx \left(ien_e[\Delta]V + \frac{1}{c}\mathbf{j}_e[\Delta]\mathbf{A}\right). \quad (18)$$

We note that in general the electron density $n_e[\Delta]$ and the current density $\mathbf{j}_e[\Delta]$ explicitly depend on the absolute value of the order parameter.

2.4 Second Order

It is convenient to introduce the following notations:

$$\dot{\theta} = 2e\Phi, \quad \mathcal{L} = \{\nabla, \mathbf{v}_s\}. \quad (19)$$

In terms of these new variables the second order correction to the action reads

$$S_2 = \frac{1}{2}\mathrm{Tr}(\mathcal{G}_0\mathcal{G}_1^{-1})^2 = -\frac{1}{8}\mathrm{tr}\left[G\dot{\theta}G\dot{\theta} + \bar{G}\dot{\theta}\bar{G}\dot{\theta} - 2F\dot{\theta}F\dot{\theta}+ \right.$$
$$\left. G\mathcal{L}G\mathcal{L} + \bar{G}\mathcal{L}\bar{G}\mathcal{L} + 2F\mathcal{L}F\mathcal{L} + 2G\dot{\theta}G\mathcal{L} - 2\bar{G}\dot{\theta}\bar{G}\mathcal{L}\right]. \quad (20)$$

Here we used the properties of the Green function (13). The form (20) of the second order correction is not quite convenient, because it contains the $\dot{\theta}L$ terms. In order to separate $\dot{\theta}$ and L we use the Ward identities (15). We write

$$G\dot{\theta}G = \theta G - G\theta + G\{\nabla, \frac{\nabla\theta}{2m}\}G - F\left(\dot{\theta} + \{\nabla, \frac{\nabla\theta}{2m}\}\right)F,$$

$$\bar{G}\dot{\theta}\bar{G} = \theta\bar{G} - \bar{G}\theta + \bar{G}\{\nabla, \frac{\nabla\theta}{2m}\}\bar{G} - F\left(\dot{\theta} - \{\nabla, \frac{\nabla\theta}{2m}\}\right)F.$$

Inserting these expressions into (20) after some simple transformations we rewrite the second order contribution as follows

$$S_2 = -\mathrm{tr}(G(\mathbf{v}_s\nabla\theta)) - \frac{1}{8}\left[GKGK + \bar{G}K\bar{G}K - GMGM + \bar{G}M\bar{G}M - G\dot{\theta}G\dot{\theta}\right.$$

$$\left. -\bar{G}\dot{\theta}\bar{G}\dot{\theta} - 2FKFK - 2F\dot{\theta}F\dot{\theta} + 2FMFM + 4F\mathcal{L}F\mathcal{L}\right]. \quad (21)$$

Here we have introduced

$$K = \{\nabla, \mathbf{u}\},$$

$$\mathbf{u} = \frac{\nabla\theta}{2m} + \mathbf{v}_s = \frac{e}{m}\left(\int_{-\infty}^{\tau} d\tau'(\nabla V(\tau')) - \frac{1}{c}\mathbf{A}\right), \qquad (22)$$

$$M = \{\nabla, \frac{\nabla\theta}{2m}\}.$$

The values $\dot\theta$ and $\mathbf{v}_s$ are now almost decoupled. The terms containing both these values were transformed into the terms containing the linear combination of these values $\mathbf{u}$ which does not depend on the phase of the order parameter field. The action (21) can be simplified further. We rewrite the first term of (21) as follows

$$-\mathrm{tr}(G(\mathbf{v}_s\nabla\theta)) = \mathrm{tr}\left[\left(m\mathbf{v}_s^2 + \frac{(\nabla\theta)^2}{4m} - m\mathbf{u}^2\right)G\right].$$

Again we decouple θ and $\mathbf{v_s}$. Making use of the identities (15) yet a couple of times we arrive at the final expression for the second order contribution to the effective action:

$$S_2 = \mathrm{tr}(m\mathbf{v}_s^2 G) - \mathrm{tr}(m\mathbf{u}^2 G) - \frac{1}{4}\mathrm{tr}(G\{\nabla, \mathbf{u}\}G\{\nabla, \mathbf{u}\})$$

$$+\frac{1}{4}\mathrm{tr}(F\{\nabla, \mathbf{u}\}F\{\nabla, \mathbf{u}\}) + \frac{1}{2}\mathrm{tr}(F\dot\theta F\dot\theta) - \frac{1}{2}\mathrm{tr}(F\{\nabla, \mathbf{v}_s\}F\{\nabla, \mathbf{v}_s\}). \qquad (23)$$

2.5   Resulting Action

Combining all contributions, we get the final result [15]

$$S = S_s[\Delta, \Phi, \mathbf{v}_s] + S_N[\Delta, V, \mathbf{A}] + S_{em}[\mathbf{E}, \mathbf{B}], \qquad (24)$$

where

$$S_s = \int dx\left(\frac{\Delta^2}{\lambda}\right) - \mathrm{Tr}\ln \mathcal{G}_0^{-1}[\Delta] + \mathrm{Tr}\ln \mathcal{G}_0^{-1}[\Delta = 0] +$$
$$\frac{1}{2}\mathrm{tr}(F\dot\theta F\dot\theta) - \frac{1}{2}\mathrm{tr}(F\{\nabla, \mathbf{v}_s\}F\{\nabla, \mathbf{v}_s\}), \qquad (25)$$

$$S_N = \int dx\left(-ie(n_e[\Delta] - n)V - \frac{1}{c}\mathbf{j}_e[\Delta]\mathbf{A} + \frac{m\mathbf{u}^2}{2}n_e[\Delta]\right)$$
$$-\frac{1}{4}\mathrm{tr}(G\{\nabla, \mathbf{u}\}G\{\nabla, \mathbf{u}\}) + \frac{1}{4}\mathrm{tr}(F\{\nabla, \mathbf{u}\}F\{\nabla, \mathbf{u}\}), \qquad (26)$$

$$S_{em} = \int dx\frac{\mathbf{E}^2 + \mathbf{B}^2}{8\pi}. \qquad (27)$$

3   Effective Action for Ultra-thin Wires

Let us now focus our attention specifically on the case of quasi-one-dimensional superconducting wires. If one assumes that deviations of the amplitude of the order parameter field from its equilibrium value are relatively small, the above effective action can be simplified further. We obtain

$$
S = \frac{s}{2} \int \frac{d\omega dq}{(2\pi)^2} \left\{ \frac{|A|^2}{Ls} + \frac{C|V|^2}{s} + \chi_E \left| qV + \frac{\omega}{c}A \right|^2 + \chi_J \left| V + \frac{i\omega}{2e}\varphi \right|^2 \right.
$$

$$
\left. + \frac{\chi_L}{4m^2} \left| iq\varphi + \frac{2e}{c}A \right|^2 + \chi_A |\delta\Delta|^2 \right\}. \tag{28}
$$

Here L and C are respectively the inductance times unit length and the capacitance per unit length of the wire. The functions χ_E, χ_J, χ_L and χ_A depend both on frequency and wave vector. They are related to the so-called polarization bubbles [15] and can be evaluated analytically for most limiting cases. Some of the expressions will be presented below.

The voltage V and the vector potential A enter the action in a quadratic form and, hence, can be exactly integrated out from the path integral. After that the effective action will only depend on φ and $\delta\Delta$. We find

$$
S = \frac{1}{2} \int \frac{d\omega dq}{(2\pi)^2} \left\{ \mathcal{F}(\omega, q)|\varphi|^2 + \chi_A |\delta\Delta|^2 \right\}, \tag{29}
$$

where

$$
\mathcal{F}(\omega, q) = \frac{\left(\frac{\chi_J}{4e^2}\omega^2 + \frac{\chi_L}{4m^2}q^2\right)\left(\frac{C}{sL} + \chi_E\left[C\omega^2 + \frac{q^2}{L}\right]\right) + \frac{\chi_J\chi_L}{4m^2}\left[C\omega^2 + \frac{q^2}{L}\right]}{\left(\frac{C}{s} + \chi_J + \chi_E q^2\right)\left(\frac{1}{sL} + \chi_E\omega^2 + \frac{e^2}{m^2}\chi_L\right) - \chi_E^2\omega^2 q^2}. \tag{30}
$$

The electromagnetic potentials are expressed as follows:

$$
V = \frac{\chi_J\left(\frac{1}{sL} + \chi_E\omega^2 + \frac{e^2}{m^2}\chi_L\right) + \frac{e^2}{m^2}\chi_E\chi_L q^2}{\left(\frac{C}{s} + \chi_J + \chi_E q^2\right)\left(\frac{1}{sL} + \chi_E\omega^2 + \frac{e^2}{m^2}\chi_L\right) - \chi_E^2\omega^2 q^2} \left(\frac{-i\omega}{2e}\varphi\right), \tag{31}
$$

$$
A = \frac{\frac{e^2}{m^2}\chi_L\left(\frac{C}{s} + \chi_J + \chi_E q^2\right) + \chi_E\chi_J\omega^2}{\left(\frac{C}{s} + \chi_J + \chi_E q^2\right)\left(\frac{1}{sL} + \chi_E\omega^2 + \frac{e^2}{m^2}\chi_L\right) - \chi_E^2\omega^2 q^2} \left(\frac{icq}{2e}\varphi\right). \tag{32}
$$

In most of the situations the wire inductance is not important and can be neglected. Therefore here and below we put $L = 0$. Then we get

$$
S = \frac{s}{2} \int \frac{d\omega dq}{(2\pi)^2} \left\{ \frac{\left(\frac{\chi_J}{4e^2}\omega^2 + \frac{\chi_L}{4m^2}q^2\right)\left(\frac{C}{s} + \chi_E q^2\right) + \frac{\chi_J\chi_L}{4m^2}q^2}{\frac{C}{s} + \chi_J + \chi_E q^2}|\varphi|^2 + \chi_A|\delta\Delta|^2 \right\}, \tag{33}
$$

and

$$V = \frac{\chi_J}{\frac{C}{s} + \chi_J + \chi_E q^2} \left(\frac{-i\omega}{2e} \varphi \right),$$ (34)

$$A = 0.$$ (35)

Here we note that the Josephson relation $V = \dot{\varphi}/2e$ holds only in the limit $\chi_J \gg C/s + \chi_E q^2$, and the latter condition holds only for low frequencies and wave vectors $Dq^2/\Delta \ll 1$, $\omega/\Delta \ll 1$.

Now let us present some expressions for the "susceptibilities" χ_E, χ_J, χ_L and χ_A which can be derived from the general results [15]. Here we will only be concerned with the practically important limit of small elastic mean free paths l. Let us define the normal state conductance of the wire σ determined by the standard Drude formula $\sigma = 2e^2 N_0 D$, where $D = v_F l/3$ is the diffusion constant. Relatively simple expressions for the χ-functions can be obtained in the limits of small and large frequencies ω and wave vectors q.

Let us start from the function χ_E. At $T \to 0$ and at small ω and q we find

$$\chi_E = \frac{\pi\sigma}{8\Delta} \left[1 - \frac{3}{8} \left(\frac{\omega}{2\Delta} \right)^2 - \frac{8}{3\pi} \frac{Dq^2}{2\Delta} \right],$$ (36)

while in the opposite large frequency limit $|\omega| \gg 2\Delta$ one gets

$$\chi_E \simeq \frac{\sigma}{|\omega| + Dq^2}.$$ (37)

It is also possible to evaluate χ_E at $\omega = 0$ and $q = 0$ and arbitrary temperatures:

$$\chi_E(0,0) = \frac{\pi\sigma}{8\Delta} \left(\tanh \frac{\Delta}{2T} - \frac{\Delta}{2T \left(\cosh \frac{\Delta}{2T} \right)^2} \right).$$ (38)

In the high temperature limit $\Delta \ll T$ we get $\chi_E(0,0) = \pi\sigma\Delta^2/96T^3$. In this limit we can use the following approximation at $\omega \neq 0$ and $q = 0$:

$$\chi_E = \frac{\sigma}{|\omega|} + \frac{2\sigma\Delta^2}{|\omega|^3} \left[\Psi \left(\frac{1}{2} \right) - \Psi \left(\frac{1}{2} + \frac{|\omega|}{2\pi T} \right) + \frac{|\omega|}{4\pi T} \Psi' \left(\frac{1}{2} + \frac{|\omega|}{2\pi T} \right) \right].$$ (39)

Here and below $\Psi(x)$ is the digamma function.

Analogous expressions can be derived for other χ-functions. For the function χ_J at low frequencies and wave vectors we have

$$\chi_J = 2e^2 N_0 \left[1 - \frac{2}{3} \left(\frac{\omega}{2\Delta} \right)^2 - \frac{\pi}{4} \frac{Dq^2}{2\Delta} \right].$$ (40)

At $\omega = 0$, $q = 0$, and arbitrary T one finds

$$\chi_J = 2\pi e^2 N_0 \Delta^2 T \sum_{\omega_\nu} \frac{1}{(\omega_\nu^2 + \Delta^2)^{3/2}} = \begin{cases} 2e^2 N_0, & T \ll \Delta, \\ \frac{7\zeta(3)}{2\pi^2} \frac{e^2 N_0 \Delta^2}{T^2}, & T \gg \Delta. \end{cases}$$ (41)

In the limit $T \gg \Delta$ and $\omega \neq 0$ we obtain

$$\chi_J = \frac{8e^2 N_0 \Delta^2}{|\omega|(\omega^2 - D^2 q^4)} \left\{ |\omega| \left[\Psi\left(\frac{1}{2} + \frac{|\omega| + Dq^2}{4\pi T}\right) - \Psi\left(\frac{1}{2}\right) \right] \right.$$

$$\left. -Dq^2 \left[\Psi\left(\frac{1}{2} + \frac{|\omega|}{2\pi T}\right) - \Psi\left(\frac{1}{2}\right) \right] \right\}. \tag{42}$$

Evaluating the function χ_L we first consider the limit $T = 0$. In this limit at low frequencies and wave vectors we get

$$\chi_L = 2\pi N_0 Dm^2 \Delta \left[1 - \frac{1}{4}\left(\frac{\omega}{2\Delta_0}\right)^2 - \frac{2}{\pi}\frac{Dq^2}{2\Delta_0} \right]. \tag{43}$$

In the limit $T \gg \Delta$ we find

$$\chi_L = \frac{4m^2 \sigma \Delta^2}{e^2(\omega^2 - D^2 q^4)} \left\{ |\omega| \left[\Psi\left(\frac{1}{2} + \frac{|\omega|}{2\pi T}\right) - \Psi\left(\frac{1}{2}\right) \right] \right.$$

$$\left. -Dq^2 \left[\Psi\left(\frac{1}{2} + \frac{|\omega| + Dq^2}{4\pi T}\right) - \Psi\left(\frac{1}{2}\right) \right] \right\}. \tag{44}$$

At zero frequency and wave vector we get $\chi_L = \pi m^2 \sigma \Delta^2 / 2e^2 T$.

Finally for the function χ_A at $T = 0$, low frequencies and wave vectors we obtain

$$\chi_A \simeq 2N_0 \left(1 + \frac{1}{3}\left(\frac{\omega}{2\Delta}\right)^2 + \frac{\pi}{4}\frac{Dq^2}{2\Delta} \right), \tag{45}$$

while in the high frequency limit we find

$$\chi_A \simeq 2N_0 \ln \frac{|\omega| + Dq^2}{\Delta}. \tag{46}$$

At temperatures $T \gg \Delta$ we get

$$\chi_A = 2N_0 \left[\ln \frac{T}{T_C} + \Psi\left(\frac{1}{2} + \frac{|\omega| + Dq^2}{4\pi T}\right) - \Psi\left(\frac{1}{2}\right) \right]. \tag{47}$$

The above expressions allow to proceed with evaluating of the QPS action practically in all interesting limiting cases.

4   QPS Action

Keeping the terms of the order q^4 and $\omega^2 q^2$ we find

$$S = \frac{s}{2} \int \frac{d\omega dq}{(2\pi)^2} \left\{ \left(\frac{C}{s}\omega^2 + \pi\sigma\Delta q^2 + \frac{\pi^2}{8}\sigma Dq^4 + \frac{\pi\sigma}{8\Delta}\omega^2 q^2 \right) \left|\frac{\varphi}{2e}\right|^2 \right.$$

$$+2N_0 \left(1 + \frac{\omega^2}{12\Delta^2} + \frac{\pi Dq^2}{8\Delta}\right) |\delta\Delta|^2 \} . \tag{48}$$

The term $\propto \omega^4$ turns out to be equal to zero.

At even smaller wave vectors, $Dq^2/2\Delta \ll 2C/\pi e^2 N_0 s \ll 1$, we get

$$S = \frac{1}{2} \int \frac{d\omega dq}{(2\pi)^2} \left\{(C\omega^2 + \pi\sigma\Delta sq^2)\left|\frac{\varphi}{2e}\right|^2 + s\chi_A|\delta\Delta|^2\right\}. \tag{49}$$

Here we have assumed $C/2e^2 N_0 s \ll 1$. This inequality is usually well satisfied for sufficiently good metals, perhaps except for the case of some specially chosen substrates. The form of the action suggests the existence of the plasma modes which can propagate along the wire. These are the so-called Mooij-Schön modes [20], the velocity of which is given by the following formula:

$$c_0 = \sqrt{\frac{\pi\sigma\Delta s}{C}}. \tag{50}$$

One can show [13] that in the case of a very long wire the action (49) yields a QPS solution described by a simple dependence $\varphi(x,\tau) = -\arctan(x/c_0\tau)$. The long time behavior of this solution results in the logarithmic interaction between two phase slips (x_1, τ_1) and (x_2, τ_2):

$$S_{\text{int}} = \frac{\mu}{2} \ln\left[\frac{(x_1 - x_2)^2 + c_0^2(\tau_1 - \tau_2)^2}{\xi^2}\right], \tag{51}$$

where

$$\mu = \frac{\pi}{4\alpha} \sqrt{\frac{sC}{4\pi\lambda_L^2}}. \tag{52}$$

Here $\alpha \simeq 1/137$ is the fine structure constant. In short wires, however, the above logarithmic interaction (51) does not play an important role and can be essentially neglected.

Now let us estimate the contribution of a single phase slip to the effective action. An approximate procedure employed here allows to obtain the correct QPS action up to a numerical prefactor of order one.

First we rewrite the action (48) in time and space domain dropping the term $\propto (Dq^2)^2$:

$$S = \frac{s}{2} \int dx\, d\tau \left\{\frac{C}{4e^2 s}\left(\frac{\partial\varphi}{\partial\tau}\right)^2 + \frac{\pi\sigma\Delta}{4e^2}\left(\frac{\partial\varphi}{\partial x}\right)^2 + \frac{\pi\sigma}{32e^2\Delta}\left(\frac{\partial^2\varphi}{\partial x\partial\tau}\right)^2\right\}$$

$$+ sN_0 \int dx\, d\tau \left\{\delta\Delta^2 + \frac{1}{12\Delta^2}\left(\frac{\partial\delta\Delta}{\partial\tau}\right)^2 + \frac{\pi D}{8\Delta}\left(\frac{\partial\delta\Delta}{\partial x}\right)^2\right\}. \tag{53}$$

Then we assume that the absolute value of the order parameter is equal to zero at time $\tau = 0$ and at the point $x = 0$. The spacial size of the QPS core is denoted as x_0, and its time duration is τ_0. The fluctuation part of the magnitude of the order parameter field $\delta\Delta(x,\tau) = \Delta(x,\tau) - \Delta_{\mathrm{BCS}}$ can be approximately expressed as follows:

$$\delta\Delta(x,\tau) = -\Delta\exp(-x^2/2x_0^2 - \tau^2/2\tau_0^2). \tag{54}$$

The QPS phase dependence on x and τ should satisfy several requirements. In a short wire and outside the QPS core the phase φ should not depend on the spatial coordinate in the zero current bias limit. On top of that, at $x = 0$ and $\tau = 0$ the phase should flip in a way to provide the change of the net phase difference across the wire by 2π. A trial function which obeys these requirements may be chosen in the following form:

$$\varphi(x,\tau) = -\frac{\pi}{2\cosh(\tau/\tau_0)}\tanh\left(\frac{x}{x_0\tanh(\tau/\tau_0)}\right). \tag{55}$$

Similar other trial functions can also be considered.

With the trial functions (54,55) the action (53) takes the form

$$S(x_0,\tau_0) = \left[a_1\frac{C}{e^2} + a_2sN_0\right]\frac{x_0}{\tau_0} + a_3sN_0D\Delta\frac{\tau_0}{x_0}$$
$$+ a_4\frac{sN_0D}{\Delta}\frac{1}{x_0\tau_0} + a_5sN_0\Delta^2x_0\tau_0 + a_6\frac{\tilde{C}}{e^2\tau_0}. \tag{56}$$

Here a_j are numerical factors of order one which depend on the precise shape of the trial functions, $\tilde{C} = CX$ is the total capacitance of the wire and X is the wire length. Here we will not discuss the capacitive effects and just neglect the last term in (56) for simplicity. As before, we also use $C/e^2N_0s \ll 1$. Minimizing the remaining action with respect to the core parameters x_0 and τ_0 we obtain

$$x_0 = \left(\frac{a_3a_4}{a_2a_5}\right)^{1/4}\sqrt{\frac{D}{\Delta}}, \quad \tau_0 = \left(\frac{a_2a_4}{a_3a_5}\right)^{1/4}\frac{1}{\Delta}. \tag{57}$$

These values provide the minimum for the QPS action, and we find

$$S_{QPS} = 2(\sqrt{a_2a_3} + \sqrt{a_4a_5})N_0s\sqrt{D\Delta}. \tag{58}$$

The results (57) and (58) are valid provided all capacitive effects can be neglected. These effects are unimportant for relatively short wires

$$X \ll \xi\frac{e^2N_0s}{C}. \tag{59}$$

Otherwise modifications to the above results should be considered.

One can also express eq. (58) in the form convenient for further comparison with experiments:

$$S_{QPS} = A\frac{R_q}{R}\frac{X}{\xi}. \tag{60}$$

Here $A = 2(\sqrt{a_2 a_3} + \sqrt{a_4 a_5})/\pi$, R is the total wire resistance, $R_q = \pi\hbar/2e^2 = 6.453$ kΩ is the resistance quantum and $\xi = \sqrt{D/\Delta}$ is the superconducting coherence length. The numerical prefactor A in the expression (60) is not known exactly. We have used several trial QPS functions, and have obtained $A \approx 1 \div 2.5$. Since our method is essentially a variational procedure, we do not expect A to deviate substantially from these values.

5 Pre-exponential Factor

The above results allow to get an idea about the suppression of QPS in ultra-thin superconducting wires depending on thickness, impurity concentration and other parameters. These results, however, are not yet sufficient to evaluate the QPS rate which has the form

$$\gamma_{QPS} = B \exp(-S_{QPS}). \tag{61}$$

The task at hand is to provide an estimate for the pre-exponential factor B in eq. (61). A general strategy to be used for this purpose is well known [23]. One can start, e.g. from the expression for the grand partition function of the wire

$$Z = \int \mathcal{D}\Delta \mathcal{D}\varphi \exp(-S) \tag{62}$$

and evaluate this path integral within the saddle point approximation. The least action paths

$$\delta S/\delta\Delta = 0, \qquad \delta S/\delta\varphi = 0 \tag{63}$$

determine all possible QPS configurations. Integrating over small fluctuations around all QPS trajectories one represents the grand partition function in terms of infinite series (each term in such series corresponds to one particular QPS saddle point). Then – at least if interaction between different quantum phase slips is small and can be neglected – one can easily sum these series and represent the final result in the form of the exponent

$$Z = \exp(-F/T), \tag{64}$$

where a formal expression for the free energy F reads

$$F = F_0 - T\sqrt{\frac{\det \delta^2 S_0/\delta Y_i \delta Y_j}{\det \delta^2 S_1/\delta Y_i \delta Y_j}} \exp(-S_{QPS}). \tag{65}$$

Here F_0 is the free energy without quantum phase slips, Y_i denote the relevant coordinates (fields), in our case $Y_1 \equiv \Delta$ and $Y_2 \equiv \varphi$, and the subscripts "0" and "1" denote the action respectively without and with one QPS.

The next step is to handle the contribution of the so-called zero modes, i.e. zero eigenvalues of the matrix $\delta^2 S_1/\delta Y_i \delta Y_j$. In our case there exist two such zero modes which correspond to uniform shifts of the QPS in space and in time. Such shifts cost no extra energy, therefore fluctuations in the corresponding directions in the functional space are non-Gaussian and require a special treatment [23]. One can demonstrate that in our case such a treatment yields the following result

$$F = F_0 - T \int_0^{1/T} d\tau \int_0^X dx \frac{\sqrt{S_{QPS}}}{\tau_0} \frac{\sqrt{S_{QPS}}}{x_0} \sqrt{\frac{\det \delta^2 S_0/\delta Y_i \delta Y_j}{\det' \delta^2 S_1/\delta Y_i \delta Y_j}} \exp(-S_{QPS}).$$

(66)

where X is the wire length and det' implies determinant with zero modes excluded.

The expression (66) is well defined and can in principle be evaluated. Note, however, that the calculation of the ratio of two determinants (66) is in general a relatively complicated problem even if the saddle point trajectories can be determined explicitly. In our case an analytical expression for the QPS trajectory is not even known and, hence, an exact evaluation of the ratio of the two determinants (66) is not possible.

Below we will conjecture that such an evaluation can only yield an unimportant numerical factor of order one in front of the pre-exponent B in (61). Hence, precise calculation of the determinants is actually not needed for our present purposes. One can support this conjecture with the aid of quite general arguments [24]. Here we provide only a simple illustration.

Consider a well known problem of a quantum particle in a periodic cosine potential. This problem is identical, e.g., to one of a Josephson junction in the quantum limit [19]. The corresponding expression for the grand partition function reads

$$Z = \int \mathcal{D}\varphi(\tau) \exp\left[-\int d\tau \left(\frac{\dot{\varphi}^2}{16E_C} + E_J(1 - \cos\varphi) \right) \right],$$

(67)

where E_C and E_J are respectively the charging and the Josephson coupling energies of a tunnel junction between two superconductors, $\varphi(\tau)$ is the Josephson phase difference across this junction. Here for simplicity we ignored dissipative terms [19,12] which can easily be restored, if necessary.

Implementing the above strategy we find a trivial and a non-trivial saddle point solutions for the action in (67), i.e. $\varphi_0 = 0$ and

$$\varphi_1(\tau, \tau_0) = 4 \arctan(\exp(\tau/\tau_0)),$$

(68)

where τ_0 plays the role of the effective QPS core size in imaginary time. Substituting (68) into the action one finds

$$S_1(\tau_0) = \frac{1}{2E_C\tau_0} + 4E_J\tau_0$$

(69)

Minimizing (69) with respect to τ_0 we reproduce the well known results

$$\tau_0 = 1/\sqrt{8E_JE_C}, \quad S_{QPS} \equiv S_1(\tau_0) = \sqrt{8E_J/E_C}.$$

(70)

Let us now consider the determinant $\det(\delta^2 S_1/\delta\varphi_1^2)$. As we have already discussed, one of the eigenvalues of this matrix is equal to zero, and the corresponding eigenfunction is known to be equal to $d\varphi_1/d\tau$. The phase fluctuations $\delta\varphi_1(\tau)$ around the instanton (68) can be expressed in the following form

$$\delta\varphi_1(\tau) = u_0 \frac{\partial}{\partial\tau}\varphi_1(\tau,\tau_0) + u_1(\tau), \qquad (71)$$

where $u_1(\tau)$ represents the contribution of all the eigenfunctions corresponding to nonzero eigenvalues of the matrix $\delta^2 S_1/\delta\varphi_1^2$. Substituting (71) into the action and calculating its variations in the absence and in the presence of QPS, δS_0 and δS_1, one finds

$$\delta S_0 \simeq \frac{32}{3} E_J \sqrt{2E_J E_C} u_0^2 + M_0, \quad \delta S_1 = M_1. \qquad (72)$$

Here M_0 and M_1 account for the contributions of all the eigenfunctions u_1 orthogonal to $d\varphi_1/d\tau$. The contribution containing u_0 in the variation δS_1 vanishes due to obvious reasons. Now let us make a crucial approximation $M_0 \simeq M_1$. Within this approximation one can set the ratio of the two determinants equal to unity and perform only one Gaussian integration over the Fourier coefficient u_0. After that one immediately finds

$$F = F_0 - T \int_0^{1/T} d\tau \frac{\gamma_{QPS}}{2}, \qquad (73)$$

where

$$\gamma_{QPS} = a \sqrt{\frac{E_J E_C}{\pi}} \left(\frac{E_J}{2E_C}\right)^{1/4} \exp\left(-\sqrt{\frac{8E_J}{E_C}}\right) \qquad (74)$$

and $a = 16/\sqrt{3} = 9.238$. Exact calculation yields the same result (74) with $a = 16$, i.e. in this case our approximation works well. One can also demonstrate that the above approximation works equally well for other situations [24].

Coming back to our problem of quantum phase slips in thin superconducting wires and employing the same approximation, from (66) we immediately arrive at eq. (73) where γ_{QPS} is now determined by the formula (61) with

$$B \approx \frac{S_{QPS} X}{\tau_0 x_0}. \qquad (75)$$

This equation provides an accurate expression for the pre-exponent B up to a numerical factor of order one. Such an accuracy is sufficient for our purposes.

6   Comparison with Experiment

Recently Bezryadin, Lau and Tinkham [17] reported a clear experimental evidence for the existence of quantum phase slips in ultra-thin (with diameters

down to 3 nm) and uniform in thickness superconducting wires. Three out of
eight samples studied in the experiments [17] stayed normal and showed no sign
of superconductivity even well below the bulk critical temperature T_C. Furthermore, in the low temperature limit the resistance of these samples was even found
to show a slight upturn with decreasing T. In view of that one can conjecture
that these samples may actually become insulating at $T \to 0$. The resistance of
other five samples [17] decreased with decreasing T. However, also for these five
samples no clear superconducting phase transition was observed.

All three non-superconducting wires (i1,i2 and i3) had the normal state resistance below the quantum unit R_q, while the normal state resistance of the
remaining five "superconducting" samples was larger than R_q. This observation
allowed the authors [17] to suggest that a dramatic difference in the behavior of
these two group of samples (otherwise having similar parameters) can be due to
the dissipative phase transition (DPT) [18,19] analogous to that observed earlier
in Josephson junctions [25].

Without going into details here, let us just point out that DPT can be observed only provided quantum phase slips can easily be created inside the wire.
The results for γ_{QPS} derived in the present paper allow to estimate a typical
average time within which one QPS event occurs in the sample. Making use of
eqs. (60), (61) and (75) we performed an estimate of such a time $t_0 = 1/\gamma_{QPS}$
for all eight samples studied in Ref. [17]. In this experiment the samples were
fabricated from $Mo_{79}Ge_{21}$ alloy. For our estimates we will use the value of the
density of states $N_0 = 1.86 \times 10^{13}$ sec/m^3 for a clean Mo, which can be extracted
from the specific heat data. The resistivity of the material was measured to be
$\rho = 1.8 \ \mu\Omega/m$, the superconducting critical temperature is $T_C = 5.5$ K. With
these numbers we estimate the coherence length to be $\xi_0 \simeq 7$ nm in agreement
with the estimate [17]. The results for t_0 are summarized in the following Table:

sample	R/d, kΩ/nm	S_0	$t_0|_{A=1}$, sec	$t_0|_{A=2}$, sec
i1	0.122	7.8	10^{-11}	10^{-8}
i2	0.110	8.7	10^{-11}	10^{-6}
i3	0.079	12.7	10^{-9}	10^{-4}
s1	0.038	25.1	10^{-4}	10^{6}
s2	0.028	33.7	1	10^{14}
s3	0.039	22.6	10^{-5}	10^{5}
ss1	0.054	15.4	10^{-7}	10^{-1}
ss2	0.044	19.6	10^{-6}	10^{2}

The action S_0 is defined by Eq. (60) with $A = 1$. The typical QPS time t_0

$$t_0 = \frac{\xi}{XAS_0\Delta} \exp(AS_0).$$

is very sensitive to the particular value of the factor A, therefore here we present
two estimates corresponding to $A = 1$ and $A = 2$.

In spite of remaining uncertainty in the prefactors some important conclusions can be drawn already from the above estimates. For instance, we observe

that the QPS rate $\gamma_{QPS} = 1/t_0$ is very high in the "insulating" wires i1, i2 and i3 for both estimates. This fact is fully consistent with the observations [17]: numerous quantum phase slips occurring in these wires completely destroy the phase coherence and, hence, superconductivity is washed out. Thus, non-superconducting behavior of these three samples is clearly due to quantum phase slips.

On the other hand, the QPS rate is notably lower for all the "superconducting" wires [17]. Possible interpretation of the experimental results for the samples s1-ss2 depends strongly on the value of A. E.g. for $A = 1$ the QPS rate is high enough practically in all samples. In this case quantum phase slips should in principle be important also for "superconducting" wires [17]. Then one can indeed relate the behavior of these samples to DPT [14], as a result of which quantum phase slips are bound in pairs and, hence, quantum fluctuations are strongly suppressed.

If, however, one chooses $A = 2$ the QPS time for the samples s1-ss2 turns out to be very long, much longer than the experimental time. Then the QPS effects should be irrelevant, and one would expect these samples to show a superconducting behavior, perhaps with the renormalized critical temperature [26]. This conclusion would also be consistent with the experimental observations [17].

Finally, let us note that all the above estimates are performed in the limit $T = 0$. This is correct if temperature is considerably below T_C. Otherwise the expression for the QPS action S_{QPS} needs to be modified.

In conclusion, we have developed a detailed microscopic theory of quantum phase slips in ultra-thin homogeneous superconducting wires. We have derived the effective QPS rate for such wires and evaluated this rate for the systems studied in recent experiments [17]. Our results are fully consistent with the experimental findings [17] which provide perhaps the first unambiguous evidence for QPS in mesoscopic metallic wires.

References

1. L.G. Aslamazov, and A.I. Larkin, Fiz. Tverd. Tela **10**, 1140 (1968) [Sov. Phys. Solid State **10**, 875 (1968)]; K. Maki, Prog. Theor. Phys. **39**, 897 (1968); R.S. Thompson, Phys. Rev. B 1, 327 (1970).
2. P.C. Hohenberg, Phys. Rev. 158, 383 (1967); N.D. Mermin and H. Wagner, Phys. Rev. Lett. 17, 1133 (1966).
3. W.A. Little, Phys. Rev. **156**, 396 (1967).
4. J.S. Langer, and V. Ambegaokar, Phys. Rev. **164**, 498 (1967); D.E. McCumber, and B.I. Halperin, Phys. Rev. B 1, 1054 (1970).
5. J.E. Lukens, R.J. Warburton, and W.W. Webb, Phys. Rev. Lett. **25**, 1180 (1970); R.S. Newbower, M.R. Beasley, and M. Tinkham, Phys. Rev. B **5**, 864 (1972).
6. A.J. van Run, J. Romijn, and J.E. Mooij, Jpn. J. Appl. Phys. **26** (1), 1765 (1987).
7. N. Giordano, Phys. Rev B **43**, 160 (1991); *ibid.*, **41**, 6350 (1990); Physica B **203**, 460 (1994) and refs. therein.
8. F. Sharifi, A.V. Herzog, and R.C. Dynes, Phys. Rev. Lett. **71**, 428 (1993). A.V. Herzog, P. Xiong, F. Sharifi, and R.C. Dynes, Phys. Rev. Lett. **76**, 668 (1996); P. Xiong, A.V. Herzog, and R.C. Dynes, Phys. Rev. Lett. **78**, 927 (1997).

9. X.S. Ling *et al.*, Phys. Rev. Lett. **74**, 805 (1995).
10. S. Saito, and Y. Murayama, Phys. Lett. A **139**, 85 (1989). Phys. Lett. A **135**, 55 (1989).
11. J.-M. Duan, Phys. Rev. Lett. **74**, 5128 (1995).
12. A.O. Caldeira, and A.J. Leggett, Phys. Rev. Lett. **46**, 211 (1981); Ann. Phys. (N.Y.) **149**, 347 (1983).
13. A.D. Zaikin, D.S. Golubev, A. van Otterlo, and G.T. Zimanyi, Phys. Rev. Lett. **78**, 1552 (1997).
14. A.D. Zaikin, D.S. Golubev, A. van Otterlo, and G.T. Zimanyi, Usp. Fiz. Nauk **168**, 244 (1998) [Physics Uspekhi **42**, 226 (1998)].
15. A. van Otterlo, D.S. Golubev, A.D. Zaikin, and G. Blatter, Eur. Phys. J. B **10**, 131 (1999).
16. F.W.J. Hekking and L.I. Glazman, Phys. Rev. B **55**, 6551, (1997).
17. A. Bezryadin, C.N. Lau, and M. Tinkham, Nature **404**, 971 (2000).
18. A. Schmid, Phys. Rev. Lett. **51**, 1506 (1983); S.A. Bulgadaev, JETP Lett. **39**, 315 (1984); F. Guinea, V. Hakim, and A. Muramatsu, Phys. Rev. Lett., **54**, 263 (1985); M.P.A. Fisher and W. Zwerger, Phys. Rev. B **32**, 6190 (1985).
19. G. Schön and A.D. Zaikin, Phys. Rep. **198**, 237 (1990).
20. J.E. Mooij and G. Schön, Phys. Rev. Lett. **55**, 114 (1985).
21. V.N. Popov, *Functional Integrals and Collective Excitations*, (Cambridge University Press, 1987).
22. H. Kleinert, Fortschr. Phys. **26**, 565 (1978).
23. See, e.g., A.I. Vainstein, V.I. Zakharov, V.A. Novikov, and M.A. Shifman, Usp. Fiz. Nauk **136**, 553 (1982) [Sov. Phys. Uspekhi **25**, 195 (1982)].
24. D.S. Golubev and A.D. Zaikin, unpublished.
25. J.S. Penttila, P.J. Hakonen, M.A. Paalanen, and E.B. Sonin, Phys. Rev. Lett. **82**, 1004 (1999).
26. Y. Oreg and A.M. Finkelstein, Phys. Rev. Lett. **83**, 191 (1999).

Part III

Molecules

Molecular Electronics:
A Review of Metal–Molecule–Metal Junctions

Jean-Philippe Bourgoin

Service de Chimie Moléculaire,CEA-Saclay,91191 Gif-sur-Yvette Cedex, France

Abstract. A review of the present state of the art in Molecular Electronics is presented. The various ways used to connect a molecule to two electrodes are first described. Then using example involving C60 molecules, it is stressed that the metal–molecule coupling governs the transport mechanism. The experimental and theoretical investigxlations on Metal–Molecule–Metal junctions based on especially designed molecules bearing clipping ends are then discussed. Finally, examples of devices are described.

1 Introduction

Molecular Electronics, understood as 'making an information processing device with a single molecule' has had very enthusiastic years in the seventies[1]. A. Aviram and M. A. Ratner in 1974 [1], and F. L. Carter [2] a bit later proposed challenging design of molecules analogous to diodes or triodes and thus in principle able to be used as building blocks of a molecular electronics. However, the interest somewhat faded away, essentially because it was too difficult at that time to connect a single molecule with any kind of electrode. About at the same time, the potential intramolecular functionality of a molecule became increasingly studied [6]. But those experiments could only be performed on ensembles of molecules. The last few years have seen a renewal of the field for two principal reasons. First, technology has evolved so that connecting a single or a few molecules was proven to be feasible. Second, the announced "ultimate limits" to be reached by the silicon technology around years 2010-2015 has prompted for the search of alternative technologies for Si technology improvement or replacement, among which Molecular Electronics is being considered. Though the soundness of this last point can be questionned - at least on this time scale - it is not unrealistic to envisage niche applications for Molecular Electronics within the domain of information processing.

The purpose of the present chapter is to review the state of the art of Molecular Electronics. The scope of this review is restricted to experiments and theory made with synthetic organic molecules, that together with nucleic Acids, Carbon nanotubes and conducting and semi-conducting nanoparticles constitute the building blocks of future molecular electronics devices.

---

[1] 'Molecular Electronics' is also used to refer to electronic transport in molecular materials. For this aspect, which is not covered in this paper, the reader is referred to ref [4,5]

Electrical transport through a single DNA fragment has recently been reported [7,8]. This opens up the possibility to take advantage of the enzymatic toolbox associated with nucleic acids to build up information processing devices [9,10], but these studies are still in their infancy.

Carbon nanotubes (NTs) on the other hand have generated a huge research activity since their discovery in 1991 [11], opening promising ways in nanotechnologies [12,13,22,15–17,43,19,27]. For electronic applications, NTs provide insulating, semiconducting or truly conducting nanoscale wires [12,13,22,15], and devices such as a junction [16] and a field-effect transistor [17,43] have even been demonstrated. With the development of methods for the controlled positioning of NTs on surfaces [27,21–25] it seems clear that a technological bottleneck is being removed and that further developpments of devices will soon follow.

Electrical transport measurements in junctions involving an individual or a few nanoparticles have been recently reported [26–28]. The results showed clear single electron tunneling effects. In order to developp these studies, the chemistry of the shell of the molecules capping the nanoparticle has to be controlled. By most aspects, this gives rise to the same problems as those of electrical transport measurements through a single or a few molecules.

Synthetic organic molecules bear the advantage that they can be synthesized with connecting groups, that in principle should help controlling the coupling of the molecule to the electrode. Besides, in the long term, once the electrical transport mechanisms involved in Metal–Molecule–Metal junctions will be understood, it will be possible to optimize the structure of the molecules on purpose.

The paper is organised as follows. First the various experimental solutions used to connect a single or a few molecules to two electrodes are described. Then results obtained with C_{60} molecules are used to introduce the role of the metal–molecule coupling. The experimental and theoretical investigations on Metal–Molecule–Metal junctions based on especially designed molecules fitted with clipping ends are then discussed. Finally, examples of devices are described.

2   Metal–Molecule–Metal Junctions

The experimental investigation of the transport properties of a single or a very few molecules contacted with two metallic electrodes has seen significant advances in the last few years. Figure 1gathers the five different configurations of electrodes used experimentally to contact molecules. The origin of these advances lays with the Scanning Tunneling Microscope(STM) which solves the connecting problem on a molecule by positioning a tip above the molecule to be connected [43]. Such an experimental setup allowed the investigation of the electronic properties of various molecules adsorbed on various conducting substrates [3,29–36,40,89,42,74,45–47,83]and eventually a single molecule [37–39]. Four categories of experiments can be defined: i) experiments where the tip is positioned over a monolayer of molecules, the actual number of molecules involved being related to the radius of curvature of the tip [3,29–32,34,35,40,89,42,74]; ii) experiments where a conjugated molecule to be measured is inserted in a matrix of molecules, like for example alkylthiols, that are less transparent to tunnel electrons; the tip is then positioned over the molecule that appear to protrude out of the matrix; this technique proved to allow the caracterisation of a single or a very few of molecules [36,45–47] iii) experiments performed by positioning the STM tip on top of individual molecules adsorbed on a substrate with a sub-monolayer density. This takes fully advantage of the imaging capability of the STM and allows to identify exactly the molecule that is measured [33,37–39] iv) experiments where a specially engineered molecule is adsorbed on top of a step of a metallic substrate and the tip of the STM is used to probe the transport along the molecule; this very elegant technique again allows to identify exactly the molecule that is being measured [48]. The conclusion, reached from these experiments and their theoretical interpretation, is that a molecule can have a non-zero conductance which strongly depends on its molecular orbitals structure [49,38,87].

Using a STM to contact a molecule, however suffers from some limitations. These include the asymmetry of the junction and a mechanical stability which might, at least at room temperature, not be good enough to allow a strong time independent coupling, like a chemical bond between the tip and the molecule. These limitations together with the improvement of lithographic techniques have led to the development of complementary techniques. In the last four years four alternatives have been proposed (see Fig. 1).

The first one concerns the use of nanoparticles encapsulated with the molecules, the electronic properties of which are investigated [50–52]. The principle of this experiment is somehow to reduce the size of the electrodes to that of a metallic cluster. It is however difficult to extract the conductance properties of a single molecule from these experiments because the law of association of conductance is no longer simple at the nanoscale [53,80]. In addition, the number of molecules involved in the measurements is not known exactly.

The second alternative makes uses of a planar configuration where the electrodes are defined by e-beam lithography at the surface of a substrate. A long molecule, like a carbon nanotube, is deposited on top of the source and drain electrode and can be influenced by the gate electrode nearby [54,55]. This type of measurements has long been restricted to molecules longer than at least 5 nm, because of the limits of nanolithography principally set up by the grain size of both the resist and metal [56]. Recent improvements of nanofabrication have however broken this 5nm limit. Bezryadin et al [57] on the one hand and Morpurgo et al [58] on the other hand have shown that starting from a 50 nm gap for example it was possible to reduce it well below 5nm by controlled sputtering and electrodeposition respectively. Park et al have induced breaking of a gold nanostructure by forcing an electromigration process to take place [59,60]eventually resulting in the formation of a typical 1 nm gap. Using this last technique, it was demonstrated recently that a Metal-C_{60}–Metal junction could be made and its transport properties at low temperature measured. Unless long molecules like NTs are investigated, this technique however lacks the certainty that a single

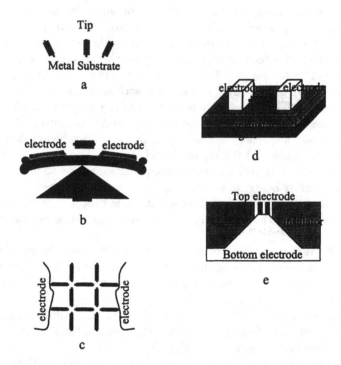

Fig. 1. Metal–Molecule(s)–Metal junctions (**a**) STM, (**b**) Mechanical breakjunction, (**c**)array of nanoclusters connected by the molecules (**d**) planar junction (**e**) nanopore approach

molecule is involved in the measurement. Indeed, the imaging of the gap is only possible with high resolution TEM and organic molecules are not apparent with such a technique.

The third alternative is the so called nanopore approach. It consists in preparing a small hole about 30nm in diameter in a silicon nitride membrane, evaporating a metal on one side of the hole thus defining a bottom electrode, adsorbing molecules on this electrode, and evaporating very carefully a top electrode [62–64,61]. This technique does definitely not allow to contact a single molecule but rather a small number (around 5000). Interesting information have nevertheless been obtained using this technique showing in particular a dependence of the transport on the molecular structure and negative differential resistance [62–64].

The fourth alternative uses a mechanical break junction to connect a molecule [66,65,45,67]. The principle of the experiment consists in breaking a small metal wire and adjusting the distance between the two facing electrodes, inherently made of the same metal, to a distance comparable to the length of the molecule. As demonstrated recently by Reed et al [65] and our group [45,67,69], this opens up the possibility of making a room temperature measurement of the transport properties of a single or a very few molecules chemically bound to two electrodes.

In particular, fabrication of mechanical break junction by e-beam lithography allows a very high stability (down to 0.2pm/hour) [70,67]. As in the case of planar electrodes, the actual number of molecules connected is unfortunately impossible to know for sure with this technique. In addition, implementation of a third electrode to influence the transport proves very problematic.

3   Metal–Molecule Coupling: Various Cases

3.1   Experiments with C_{60}

C_{60} molecules are about 7Å in diameter. Since their discovery in 1985 they have been studied by a number of techniques including by STM after adsorption on a substrate. Depending on the strength of the electronic coupling between the substrate and the molecule, the transport mechanism proves very different. In a series of experiment, Porath et al used a substrate consisting of a metal covered with a thin insulating layer. [33,90]. They recorded STM images and performed scanning tunneling spectroscopy on an isolated C_{60} molecule. Typical IV curves are shown in Figure 2 together with theoretical curves that reproduce the observed experimental behavior. The interpretation of these experimental results considers that the C_{60} molecule is embedded in a double tunnel junction configuration. The authors used a model adapted from the single electron transport orthodox theory to take into account the discrete molecular energy spectrum. By construction the C_{60} molecule is here in a configuration of low metal–molecule coupling strength.

The situation can be very different if no insulating layer is deposited on the substrate and the C_{60} molecule is directly adsorbed on a metallic substrate. This configuration was tested in refs. [37,38]. A tungsten tip was positioned on top of a C_{60} molecule adsorbed on a Au (110) surface. The recorded IV characteristics were linear at low bias voltages as shown in Fig.3. At low bias voltage a tip apex-C_{60}-Au(110) junction resistance of 54.8 MΩ was measured. These experiments were interpreted as follows. On the contrary to the abovementionned experiments where the coupling between the molecule and the metal is negligible, in the present case, the direct adsorption of the C_{60} molecule on the gold surface gives rise to a situation where i) the coupling is strong enough to ensure that a tunneling current can flow coherently through the junction, but ii) remains weak enough to preserve the electronic identity of the molecule. The linear IVs are a consequence of the absence of a molecular-level resonance in the energy window probed and of the broadening of the molecular levels by interaction with the metal surface. As will be discussed in the Molecular Devices Section, these authors have shown that an electromechanical device based on this transport mechanism can be made.

A naive conclusion that can be reached from these two sets of experiments is that it suffice to adsorb directly C_{60} molecules on a gold surface and connect them with a tip to ensure a coupling that is strong enough to give rise to a tunneling transport regime in the junction.

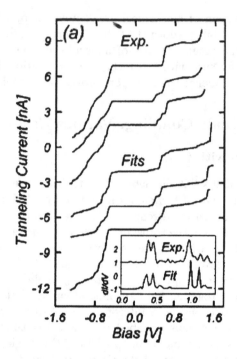

Fig. 2. Example of I-V curves recorded with an STM tip positionned above a C_{60} molecule. From ref [90]

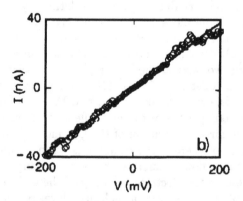

Fig. 3. Example of IV curves recorded with an STM tip positioned above a C60 molecule adsorbed on gold. From ref [38]

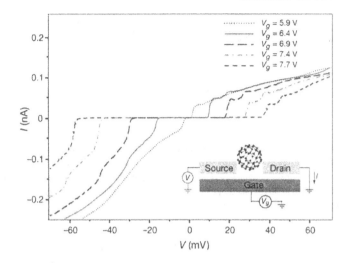

Fig. 4. Example of I-V curves recorded with a C_{60} molecule inserted in a nanojunction prepared by electromigration. From ref [60]

A very recent experiment by Park et al where a C_{60} molecule (or a very few of them) has been connected by two electrodes shows however that this is not the case. This experiment is indeed the first experimental example of a single molecule transistor. A nanometer size gap was prepared by the controlled electromigration technique described above [59,60]. C_{60} molecules were adsorbed in the gap and IVs were recorded. TThey sshowed a highly non linear behavior with current suppression at low bias as shown in Fig.4. These results seems to prove unambiguously that single electron transport occurs in the junction. This is the signature of a situation of weak coupling between the C_{60} molecule and the electrodes.

These examples demonstrate that the control of the strength of the coupling between C_{60} molecules and a substrate is far from being experimentally mastered.

3.2 Experimental Investigation on Molecules Functionnalized with Connecting Ends

The molecules that are mostly investigated for Molecular Electronics applications usually consist of a central part and one, or two (or more) clipping ends (see Fig. 5). The central part of the molecule is usually conjugated in order to facilitate the transport through the junction. In an increasing number of studies, the central part of the molecule is engineered to favor a given property like for example asymmetry of the structure [40,62,65,72,73]. The clipping ends are supposed to practically solve the problem of the connection between the electrodes and the molecule because they allow the molecule to be self-assembled onto the electrodes surfaces. In addition, this should result in a controlled coupling be-

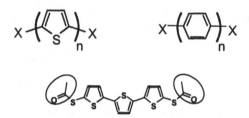

Fig. 5. Example of molecules with reactive groups at the end (X= S, Se, Te, NC). The molecules are shown like they are involved in the junctions except for the bottom one. The circled groups on the bottom molecule(refered to as terthiophendithiol in the text) are protecting groups that are removed right before the experiment. The top left molecule with n=1 and X=S is the benzenedithiol molecule.

tween the electrode and the molecule. So far, molecules with clipping ends like S, Se, Te, NC group, or electron rich conjugated group have been synthesized [48,72,73]. Among those molecules, some have been experimentally investigated in Metal–Molecule–Metal junction.

STM on molecules with π−conjugated connecting groups One experiment has been reported where molecules with fluoranthene groups at the ends have been adsorbed on the double atomic steps of a copper (100) surface [48]. The central conjugated part of the molecule is "isolated" from the bottom copper terrace by 3,5-di-tert-butylphenyl substituents used as spacers. A STM tip was then used as a type of tunneling potentiometer. From cross-sectional profiles of the apparent height of the molecular wire along its length, quantitative value of the intramolecular tunnel barrier could be determined in agreement with theory. In this case, the authors concluded that the fluoranthene groups provided a strong coupling to the upper terrace of copper and that the tranport occured through virtual resonance tunneling, i.e. through nonresonant tunneling through the tails(resulting from the electronic coupling to the metallic electrodes via the fluoranthene group) of the HOMO-LUMO manifold close to the Fermi level.

STM on thiol terminated molecules A few experiments have been performed by positioning a STM tip on top of molecules fitted with a thiol clipping group at one end [36,42,47,44] or at both ends [75,41,45,74], adsorbed on a metallic substrate. From these experiments, a few conclusions were drawn. The first one concerns the enhanced conductivity of the conjugated molecules compared to the saturated ones. [36,44,45,47]. The second conclusion is that at large bias ($\geq$ 200 mV typically), the molecular energy levels are dominating the transport [42,75,41]. The study of the low bias voltage range is not emphasized in most papers. In a few papers however, the zero bias conductance was observed to be non zero, and the conductance remained constant to the first order at low bias [42,47,75].

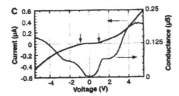

Fig. 6. I-V curves recorded on a metal–benzenedithiol–metal junction realized in a Mechanically Controllable Breakjunction (from ref. [65])

Concerning the transport at higher bias, both symmetrical and asymmetrical IVs were recorded, depending on the nature of the molecule investigated. In refs [42,74] the molecules had only a thiol group at one end. the corresponding IVs were asymmetrical. This behavior is expected to show up if the metal–molecule couplings are different at both ends of the molecule. Conversely, the IVs recorded on a xylyldithiol molecule by Tian et al [75,41], are symmetrical. The authors interpreted this results as arising i) from a symmetrical coupling at both ends of molecule and ii) from the molecular energy levels floating consequently midway between the electrochemical potential of the electrodes. In [89,41], a transport model based on a scattering mechanism was used, assuming that the coupling was strong (see below).

Mechanical Breakjunction investigation on dithiolmolecules So far, the transport properties of two types of molecules have been measured by insertion in the gap of a MCB: benzendithiol and terthiophenedithiol. Fig. 6 shows IVs recorded on a MCB in the gap of which a benzenedithiol molecule was inserted [65]. The experimental IVs were symmetrical with an apparent current suppression at low bias and steps in the conductance vs V curve. These results were initially interpreted as resulting from Coulomb Blockade, i.e. the authors assumed a weak coupling between the molecule and the metal, or in other terms that the Au-S-C bonds corresponds to a tunnel junction. Other interpretations have been put forward since then [88,75,76,78], considering a scattering mechanism involving the molecular energy levels: a common assumption (except in ref [76]) is that the molecule metal coupling is strong. A conservative conclusion about this experiment is that the molecule can be symmetrically coupled to two electrodes.

Fig 7 shows the two types of curves that were recorded on MCBs in the gap of which terthiophenedithiol molecules were adsorbed [45,67,69]. Asymmetric I-Vs of type (**a**) were more often observed and more stable than symmetric ones of type (**b**). The measured zero bias conductance of type (**a**) junctions was of the order of 10 nS. The asymmetric I-V characteristic was non-linear with step-like features, and the current increased linearly at large bias voltage. The measured zero bias conductance of type (**b**) junctions was larger, about 80 nS. The symmetric I-V characteristic was also non-linear with smaller step-like

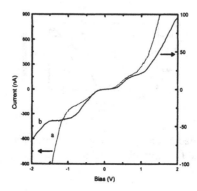

Fig. 7. I-V curves recorded on a metal–terthiophenedithiol–metal junction realized in a Mechanically Controllable Breakjunction (from ref. [67])

features. At $V \geq 1$ V, the current rose faster than linearly with V. These results were interpreted by considering explicitly the discreteness of the molecular energy spectrum [45,67,69]. A scattering model was used at low bias to interpret the experimental results, whereas at higher bias, the results were shown to be accounted for qualitatively both by a scattering and a sequential tunneling model. The validity of the two models depends on the strength of coupling going with the Au-S-C link. A question that these experimental results leave open.

3.3 Modelling of the Transport in Metal–Molecule–Metal Junctions

The theory of metal–insulator–metal junctions has been given a strong push because of the need for interpretation of STM images. The theoretical models developed towards this goal have been reviewed for example in ref. [81,82]. Though part of these theories can be adapted to the case where the insulator is an adsorbed molecule, I will not try to cover this topic extensively. Here I will rather mention only the models that have been explicitly developed to account for the transport through metal–molecule–metal junctions. I will not discuss them in detail but present their principal characteristics, findings and important parameters.

These models can be classified in two extreme cases depending on the assumption made on the coupling.

In a first case, one assumes that the metal–molecule–metal coupling is small (*i.e.* the conductance of the contact is smaller than e^2/h) at both ends of the molecule and that the energies involved in the tunneling processes are large compared to kT. In addition, the broadening of the levels (that the electron will populate by tunneling) due to interactions with molecules in the surrounding or molecular vibrations has to be larger than the broadening due to the coupling to the electrodes. Then the transport occurs through sequential tunneling. The electron tunnels first through say the left contact in the molecule and finally tunnels through the right contact. Models have been developed based on the orthodox theory of Coulomb Blockade, taking into account the molecular energy levels [90,67]. Such a model was for example proven to account properly for the experiments performed on a C_{60} molecule adsorbed on a thin insulator (see subsection -'experiment with C_{60}above'). It was also shown to account qualitatively for the measurements performed on a metal–terthiophenedithiol–metal junction [67]. The models developed so far have used a rough estimate of the energy difference between different charged states of the molecule based on one electron orbitals of the molecule considered as isolated from the metallic electrodes. This description definitely needs improvement. In these models, the key parameters are: the description of the coupling and the position of the molecular energy levels compared to the Fermi levels of the electrodes, because it changes the transition energy between different charged states of the molecule. It should be noted that at low temperature, the sequential model predicts a zero conductance at low bias.

In the second case one considers that the coupling is strong between the molecule and the metallic leads and calculates the transport through the junction using the Landauer theory which relates the current, from one side of the junction to the other one, to the transmission probability for an electron to scatter through the molecule via the finite voltage, finite temperature Landauer formula [68].

Until recently, the molecular description has always been at the semi-empirical model and more precisely with the extended Hückel (EH) model. In this model, the transmission function $T(E)$ (where E is the energy of the incoming electron) of the junction is calculated. As can be seen on the example shown in Fig. 8, the transmission function shows i) some resonances corresponding to a molecular level coming into resonance with the injecting energy and ii) lower transmission regions in between the resonances where the transport occurs by coherent tunneling trough the tails of the molecular levels broadened by the coupling to the metallic electrodes.

The zero bias conductance is obtained from the Landauer formula:

$$G = \frac{2e^2}{h} T(E_F) \tag{1}$$

In the linear regime, the conductance can be expressed as

$$G = G_0 e^{-\gamma L} \tag{2}$$

where G_0 is a function of the coupling strength at both ends of the molecule, and γ is a function of the chemical structure of the molecule of length L. Both G_0 and γ depend on the position of the molecular energy scale compared to the Fermi level of the electrodes. A lot of studies have been devoted to the influence of the molecular design on the decay length γ [88,75,41,49,84,85,87]. Recently, some emphasis has been put on the influence of the coupling strength on the conductance [88,69,78,80]. The surface geometry (on top compared to hollow site adsorption of the sulfur atom onto gold for example, (100) vs (111)

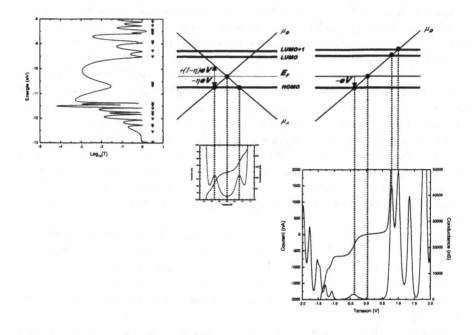

Fig. 8. left: Transmission spectrum T(E) calculated for the terthiophenedithiol molecule[67]. middle: I(V) and dI/d(V)(V) calculated using (3) assuming a symmetrical coupling at both ends of the molecule. right: I(V) and dI/d(V)(V) calculated assuming that the molecule is strongly coupled to one electrode and not to the other

orientation of the gold clusters terminating the electrodes) is slightly influencing the conductance while the sulfur-gold bond distance influence is much more important. In most cases (except in ref [78]) the calculated values of conductances are larger than the experimental values which points to the fact that the molecule–metal coupling strength is probably overestimated compared to what it is in reality. In addition, the theoretical predictions drastically depend on the energy of the electron injected in the junction. It seems accepted now that the HOMO of a molecule chemisorbed on gold through a sulfur or selenium link is tied to the Fermi level of the electrodes and not to the vacuum, and that the HOMO of the benzenedithiol, xylyldithiol and terthiophenedithiol lie closer to the Fermi level of the electrodes than the LUMO [41,67,78,79]. The experimental investigation that bring confirmations of this point are scarce however. They are based on the photoemission spectroscopy performed on self-assembled monolayers (SAM) of these molecules onto a metallic substrate. To our knowledge, no direct study of benzenedithiol or xylyldithiol have been reported. In a preliminary experiment on terthiophenedithiol SAMs on gold, we have shown that $E_F - E_{HOMO} = 0.5 \pm 0.1 eV$. [86] This value is however only an estimate since in a metal–molecule–metal junction, the molecule is adsorbed at both ends giving rise to a metal–molecule charge transfer at both ends. The additionnal charge transfer compared to a SAM on a flat surface likely changes $E_F - E_{HOMO}$. In this model, the current at larger bias is obtained according to

$$I(V) = \frac{e}{\pi\hbar} \int_{-\infty}^{+\infty} T(E,V)\left(f(E - \mu_1) - f(E - \mu_2)\right) dE \qquad (3)$$

in which the effect of the bias voltage V on the $T(E)$ spectrum is included; $\mu_{1,2}$ is the chemical potential of the electrode 1,2, and $f(\epsilon)$ denotes the Fermi function at the temperature of the experiment. In the case of a symmetric junction, the voltage at the molecule lies halfway between the electrode voltages. In order to calculate the current, the crude approximation usually made is that $T(E,V) \simeq T(E - eV/2, 0)$ where $T(E, 0)$ is the spectrum calculated in the absence of any bias. Equation 3 then predicts the I-V characteristic shown in Fig. 8 in the case of terthiophenedithiol. For both polarities, the first resonances occur through the HOMO and the HOMO-1 levels. This results from the assumption that the equilibrium Fermi level is closer to the HOMO than to the LUMO [41]. In this model, the shape and height of the steps is only due to the broadening of the molecular levels by the coupling with the electrodes. In particular, the height of a step in the I-V curve is directly proportional to the area of the corresponding peak in the $T(E,V)$ spectrum. The experimental curve is well reproduced for the $[-0.5V, 0.5V]$ range. Outside of this range, the calculated current is higher than the experimental one. It should be noted that the only adjustable parameter used in the calculation is the position of the Fermi level, which is again a key parameter. In the case of Fig. 8 $E_F - E_{HOMO} = 0.4$ eV [67] was used. In this calculation, the magnitude of the current depends on the following points: i) the overlap between the orbitals of the outer Au atoms and of the sulfur atoms; ii) the localisation of the molecular orbital and its symmetry, iii) perfect symmetry between the two contacts. The applied voltage is expected, especially at high bias, to affect points ii) and iii), both of which should tend to decrease the calculated current and thus reduce the discrepancy.

Recently the theoretical description of transport in a Metal–Molecule–Metal junction has been improved along three directions.

First Yaliraki et al extended their previous models to treat the molecular part and the contact at an ab-initio level and compared the results with those obtained with the EH description for xylyldithiol and benzenedithiol [79]. The trends observed with the EH models concerning the influence of the gold-sulfur bond length and on top vs hollow site sulfur adsorption site, are confirmed by the ab-initio models with two main differences: the conductance values are lower in the latter case and the gold sulfur charge transfer inluence is much larger. In addition, this work confirmed that the HOMO level is closer to Fermi level of the electrodes than the LUMO. This technique however suffers from the known poor description of the molecular affinity levels. In this paper the authors pointed out that if in the ground state, the molecule contains N electrons, the transmission problem is a N+1 electron problem. Thus the electron population in the molecule must be determined self consistently under an applied bias.

This is the approach that Di ventra et al followed [76]. They reported first principles calculations of the I-V characteristics of a molecular junction. This constitute a second important recent development of the field. The metal in that case is described by a Jellium model with an electron density taken equal to that of metallic gold.The I-V characteristics were calculated by first computing the electron wave functions by solving the Lippman-Schwinger equation iteratively to self-consistency in the steady state.The current is then calculated from the wave functions of the electrode-molecule system. The experimental curve was reproduced qualitatively. Two discrepancies appear nevertheless. The first one concerns the position of the peaks in the conductance plot at 2.4 V in the calculated curve compared to ca. 1V in the experience. This is explained by the authors as a arising from known limitations of the local density approximation they used. The second discrepancy comes from the much larger current predicted by the calculation compared to the experience. This discrepancy is reduced when the gold sulfur contact is occuring through a single gold atom. However, even in that case, the calculation still predicts a current about one order of magnitude higher than the observed one. This points to the fact that the coupling in this calculation is still overestimated. On the other hand, the authors observed that the sulfur -gold contact has a very low density of states and that this indeed correspond to "a barrier through which the electron must tunnel" [76], a fact that was also shown to be very likely in terthiophendithiol molecules [92]. If the actual coupling strength is lower than the one that gave rise to this conclusion, then the actual barrier through which the electron has to tunnel is likely even less transparent. This brings back to conditions where the transport may be sequential rather than coherent. The authors have adressed this point in ref. [77] and calculated the delay time an electron spends in the scattering region.They found $2fs$ a time too short for charge effects to take place. This may not be a conclusive argument however since the definition of the tunneling time is still quite an open question [93,99].

The issue of the tunneling time is also a central one in recent developments considering inelastic effects on electron transport through metal–molecule–metal junctions, which is a third direction in which the modelling of the transport has been recently improved [97,96,94,95]. In ref [97] it is stressed that the transport behavior depends on the ratio between the electronic and lattice time scales. When the tunneling time is shorter than the lattice vibration time, it is shown that the transmission is very similar to the rigid like case, which is the approximation made in the references cited except refs [97,96,94,95]. Conversely when the lattice vibration frequency increases, the transport becomes polaron like. In ref [94,95], the transition between the two regimes occurs with the increase in the length of the molecule. In the case of short molecules (involving less than 10 atoms along the backbone), the electron has no time to couple efficiently with the vibrations and thus the e^--phonons effects can be neglected. For longer molecules this is not the case, and it is shown that the inelastic effects can significantly increase the tunneling current. These studies are however still restricted to model molecules and did not attempt to compare with the experiments. Studies in this direction will very likely developp in the near future since i) various inelastic effects have been observed in experiments with STM on adsorbed molecules on metallic substrates (see ref. [103])and ii)recent calculations of the tunneling time through a molecule show that it is in the same range as intramolecular vibrations [99].

An additionnal point is the following: for the experiments that are not made under a high vacuum, the molecule in the junction is very likely surrounded by some gas (air, argon, ..) and solvent molecules (remaining from the preparation for example). This is not taken into account in the theoretical models mentionned above. Indeed, a smoothing of the theoretical I(V) curves of these models is expected due to this polarizable surrounding. This remark connects metal–molecule–metal junctions experiments to recent electrochemical STM measurements on single adsorbed molecules [30] and to experiments measuring the electron transfer rates in bridged molecular systems [100]. On the basis of these experiments, a debate between coherent tunneling and sequential tunneling shows up in a paper by Han et al [30].This point is also extensively discussed by Segal et al [100]. It should be noted, however, that the sequential model derived in that paper is a sequential hopping model that neglects the charging effects on the dynamics of charge carriers.

Summarizing, the transport in metal–molecule–metal junctions may be approached by two extremes: that of sequential tunneling (weak metal–molecule couplings) and that of coherent tunneling (strong couplings). Two questions arise: "is it possible to find independent estimates of the strength of coupling?" and "how likely are the models?".

Spectroscopy gives a positive answer to the first question as shown recently by Vondrak et al [98]. In this study the interfacial electronic structure of SAMS of pentafluorothiophenolate on Copper(111) has been investigated. It was shown that the HOMO-LUMO gap is strongly reduced by adsorption of the molecule on the copper surface and that this seems to originate from a strong Cu-S coupling. To our knowledge, these experiments have not yet been performed with "electrically investigated" molecules adsorbed on gold.

Considering the second question, the situation is as follows. At low bias, the non zero conductance that has been measured in many experiments, seems accounted for by the scattering theory. At larger bias, in the case of asymmetric couplings, with one end of the molecule very strongly coupled to the electrode and the other end only weakly coupled, the coherent model can be applied because the problem is then typically reduced to that of a simple tunneling problem. At larger bias in the case of symmetric couplings, the situation is more delicate as soon as one molecular level comes in resonance with the Fermi level of one of the electrodes. At that energy, the transmission probability is then 1. Thus any electron impinging on the junction will be transmitted and the average time of occupation of the molecular level will rise, thus likely limiting the validity of the coherent model. At least, the coherent model should be calculated self-consistently to take into account the fact that the molecule is charged.

4 Molecular Devices

Beside the various devices that have been developed with Carbon nanotubes (see the introduction), there exist to date to the author's knowledge, only two demonstration of three terminal devices based on a single molecule.

The first one is based on the electromechanical modulation of the transport through a C_{60} molecule adsorbed on a surface and addressed with a STM tip [37]. The principle of the modulation consists in varying the tip C_{60} distance, from a position where the C_{60} geometry is relaxed to a position where it is constrained. In the latter case, the current is much higher (three orders of magnitude) than in the former one. Based on this example, an integrated gate based on C_{60} electromechanical devices was proposed but has not been yet realized [101]. It can be noticed that the principle of electromechanical induced variation of the current carried through a Metal–Molecule–Metal junction has also been demonstrated in Mechanical breakjunctions [67].

The other example is also using a C_{60} molecule but the molecule is embedded in a planar junction in that case, and the modulation is electrostatic [60]. The device and results of the gate modulation are shown in Fig. 4. The transport mechanism has been described in section "experiments with C_{60}" above.

Finally, it should be noticed that Di ventra et al on the one hand and Emberly and Kirczenow on the other hand recently reported on calculations showing that the electrostatic modulation of the current carried coherently through a metal–(benzenedithiol or-polyacetylenedithiol)–metal junction should give rise to sizeable effects [77,104].

5 Conclusion

During the last few years, impressive advances have been made in the experimental realization of Metal–Molecule–Metal junctions. This allowed to experimentally investigate the transport mechanisms and to fabricate the first molecular devices.

The strength of coupling between the molecule and the metal electrode is a key parameter that governs the mechanism of transport. If the coupling is weak, then the transport is dominantly sequential. Conversely, if the coupling is strong then the transport mechanism is dominantly coherent. Some experimental demonstrations of these two regimes have been made. The transport regime in the experimentally relevant case of molecules adsorbed onto metallic electrodes through a sulfur link, remains however unclear.

Whatever the transport regime, a key parameter, that is also related to the strength of coupling, is the position of the molecular energy levels compared to the Fermi level of the electrodes. This parameter governs the amount of current carried through the junction.

Finally, one should note that the elucidation of the transport mechanisms in Metal–Molecule–Metal junctions is obviously a bottleneck for the future development of information processing architectures based on these molecular junctions. In particular, sequential transport leads to solutions conceived within the frame of single electronics, while coherent transport requires the development of a molecular state engineering for the transport occurring off resonance and a waveguide approach for the transport occuring at resonance.

I gratefully thank D. Esteve and M. Devoret for enlightening discussions.

References

1. A. Aviram, M. A. Ratner: Chem. Phys. Lett. **29**, 277 (1974).
2. F. L. Carter: 2nd Intl Symp. Molecular Electronic Devices (M. Dekker, New York) 149 (1982).
3. G. Lambin, M. H. Delvaux, A. Calderone, R. Lazzaroni, J. L. Bredas, T. C. Clarke, J. P. Rabe: Mol. Cryst. Liq. Cryst. 235, 75 (1993).
4. J. Jortner, M. A.Ratner: *Molecular Electronics*(Blackwell, London, 1997)
5. W.R. Salaneck, K. Seki, A. Kahn, J.J. Pireaux: *Conjugated polymers and molecular interfaces* (Dekker, NewYork)
6. P. F. Barbara, T. J. Meyer, M. A. Ratner: J. Phys. Chem. **100**, 13148 (1996)
7. D. Porath, A. Bezryadin, S. de Vries, et al.: Nature **403**, 635 (2000)
8. H. W. Fink, C. Schonenberger: Nature **398**, 407 (1999)
9. E. Braun, Y. Eichen, U. Sivan, et al.: Nature **391**, 775 (1998)
10. C. A. Mirkin, R. L. Letsinger, R. C. Mucic, et al.: Nature **382**, 607 (1996)
11. S. Iijima: Nature **354** 56 (1991)
12. W.A. de Heer, J.-M. Bonard, K. Fauth, A. Châtelain, L. Forro, D.Ugarte: Adv. Mater. **9** 87 (1997)
13. J. W. Mintmire, B. I. Dunlap, C. T. White: Phys. Rev.Lett. **68**, 631 (1992) , b) N. Hamada, S. Sawada, A. Oshiyama, ibid., p. 1579, c) M. S. Dresselhaus, G. Dresselhaus, P. C. Eklund: *Science of Fullerenes and Carbon Nanotubes* (Academic Press, New York, 1996)
14. S.J. Tans, M.H. Devoret, H. Dai, A. Thess, R.E. Smalley, L.J. Geerligs, C. Dekker: Nature **386**, 474 (1997) b) M. Bockrath, D.H. Cobden, P.L. McEuen, N.G. Chopra, A. Zettl, A. Thess, R.E. Smalley: Science **275**, 1922 (1997)
15. P. G. Collins, A. Zettl, H. Bando, A. Thess, R. E. Smalley: Science **278**, 100(1997)
16. R. D. Antonov, A. T. Johnson: Phys. Rev. Lett **83**, 3274 (1999) b) Z. Yao, H. W. C. Postma, L. Balents, C. Dekker: Nature **402**, 273 (1999) c)M. S. Fuhrer, J. Nygard, L. Shih et al.: Science **288**, (2000) 494
17. a) S.J. Tans, A.R.M. Vershueren, C. Dekker: Nature **39**, 49 (1998)b) R. Martel, T. Schmidt, T. Hertel, P. Avouris: Appl.Phys. Lett. **73**, 2447 (1998). c) L. Roschier, J. Penttilae, M. Martin, P. Hakonen, M. Paalanen, U. Tapper, E. Kauppinen, C. Journet, P. Bernier: Appl.Phys. Lett. **75**, 728 (1999).
18. H.T. Soh, A.F. Morpurgo, J. Kong, C.M. Marcus, C.F. Quate, H. Dai: Appl. Phys. Lett. **75**, 627 (1999)
19. H. Dai, J.H. Hafner, A.G. Rinzler, D.T. Colbert, R.E. Smalley: Nature **384**, 147,(1996)
20. P. Kim and C. M. Lieber Science **286**, 2148 (1999)
21. J. Kong, H.T. Soh, A.M. Cassell, C.F. Quate, H. Dai: Nature **39**, 878 (1998)
22. Z.F. Ren, Z.P. Huang, D.Z. Wang et al.: Appl. Phys. Lett **75**, 1086 (1999)
23. J. Liu, M.J. Casavant, M. Cox et al.: Chem. Phys. Lett. **303**, 125 (1999)
24. S. Gerdes, T. Ondarcuhu, S. Cholet, et al., Europhysics Letters **48**, 292 (1999)
25. K.H. Choi, J.P. Bourgoin, S. Auvray, D. Esteve, G. S. Duesberg, S. Roth, M. Burghard: Surf. Sci. **462** 195(2000)

26. D. L. Klein, R. Roth, A. K. L. Lim, et al.: Nature **389**, 699 (1997)
27. S. H. M. Persson, L. Olofsson, and L. Gunnarsson: Appl. Phys. Lett. **74**, 2546 (1999)
28. M. Dorogi, J. Gomez, R. Osifchin, et al.: Phys. Rev. B **52**, 9071 (1995)
29. M. Cyr, B. Venkataraman, G. W. Flynn, A. Black, G. M. Whitesides: J. Phys. Chem. **100**, 13747 (1996)
30. W. Han, E. N. Durantini, T. A. Moore et al.: J. Phys. Chem. B **101**, 10719 (1997)
31. H. Nejoh: Nature **353**, 640 (1991)
32. W. Mizutani, M. Shigeno, K. Kajimura, M. Ono: Ultramicroscopy **42-44**, 236 (1991)
33. D. Porath, O. Millo: J. Appl. Phys. **81**, 2241 (1997)
34. B. Michel, G. Travaglini, H. Rohrer, C. Joachim, M. Amrein: Z. Phys. B **76**, 99 (1989)
35. X. Lu, K. W. Hipps, X. D. Wang, U. Marzur: J. Am. Chem. Soc. **118**, 7197 (1996)
36. L. A. Bumm, J. J. Arnold, M. T. Cygan et al.: Science **271**, 1705 (1996)
37. C. Joachim, J. K. Gimzewski, R. R. Schlitter, C. Chavy: Phys. Rev. Lett. **74**, 2102 (1995)
38. C. Joachim, J. K. Gimzewski: Europhys. Lett. **30**, 409 (1995)
39. B. C. Stipe, M. A. Rezaei, W. Ho: Science **280**, 1732 (1998)
40. R. M. Metzger, B. Chen, U. Höpfner et al.: J. Am. Chem. Soc. **119**, 10455 (1997)
41. S. Datta, W. Tian, S. Hong et al.: Phys. Rev. Lett. **79**, 2530 (1997)
42. A. Dhirani, P.-H. Lin, P. Guyot-Sionnest, R. W. Zehner, L. R. Sita: J. Chem. Phys. **106**, 5249 (1997)
43. A. Aviram, C. Joachim, M. Pomerantz: Chem. Phys. Lett. 146, 490 (1988)
44. L. A. Bumm, J. J. Arnold, T. D. Dunbar, et al.: J. Phys. Chem. B **103**, 8122 (1999)
45. C. Kergueris, J. P. Bourgoin, and S. Palacin: Nanotechnology 10, 8 (1999)
46. J. Chen, M. A. Reed, C. L. Asplund, et al.: Appl. Phys. Lett. **75**, 624 (1999)
47. G. Leatherman, E. N. Durantini, D. Gust, et al.: J. Phys. Chem. B **103**, 4006 (1999)
48. V. J. Langlais, R. R. Schlittler, H. Tang, et al.: Phys. Rev. Lett. **83**, 2809 (1999)
49. P. Sautet and C. Joachim: Chem. Phys. Lett. **185**, 23 (1989)
50. R. P. Andres, J. D. Bielefeld, J. I. Henderson, et al.: Science **273**, 1690 (1996)
51. J. P. Bourgoin, C. Kergueris, E. Lefevre, et al.: Thin Solid Films **327-329**, 515 (1998)
52. W. P. McConnell, J. P. Novak, Louis C. Brousseau III et al.: J. Phys. Chem. B **104**, 8925 (2000)
53. M. Magoga, C. Joachim: Phys. Rev. B **59**, 16011 (1999)
54. S. J. Tans, M. H. Devoret, H. Dai, A. Thess, R. E. Smalley, L. J. Geerligs, C. Dekker: Nature. **386**, 474 (1997)
55. T. W. Ebbesen, H. J. Lezec, H. Hiura, J. W. Bennett, H. F. Ghaemi, T. Thio: Nature. **382**, 54 (1996)
56. V. Rousset, C. Joachim, B. Rousset, N. Fabre: J. Phys.III, **5**, 1985 (1995)
57. A. Bezryadin and C. Dekker: J.Vac. Sci. Technol. B **15**, 793 (1997)
58. A. F. Morpurgo, C. M. Marcus, and D. B. Robinson: Appl. Phys. Lett. **74**, 2084 (1999)
59. H. Park, A. K. L. Lim, A. P. Alivisatos, et al.: Appl. Phys. Lett. **75**, 301 (1999)
60. H. Park, J. Park, A. K. L. Lim, et al.: Nature **407**, 57 (2000)
61. J. Chen, L. C. Calvet, M. A. Reed, et al.: Chem. Phys. Lett. **313**, 741 (1999)
62. J. Chen, M. A. Reed, A. M. Rawlett, et al.: Science **286**, 1550 (1999)

63. J. Chen, W. Wang, M. A. Reed, et al.: Appl. Phys. Lett. **77**, 1224 (2000)
64. C. Zhou, M. R. Deshpande, M. A. Reed, et al.: Appl. Phys. Lett. **71**, 611 (1997)
65. M. A. Reed et al.: Science **278**, 252 (1997)
66. C. J. Muller, B.J. Vleeming, M. A. Reed et al.: Nanotechnology **7**, 409 (1996)
67. C. Kergueris, J. P. Bourgoin, S. Palacin, et al.: Phys. Rev. B **59**, 12505 (1999)
68. M. Büttiker, Y. Imry, R. Landauer, S. Pinhas: Phys. Rev. B **31**, 6207 (1985)
69. C Kergueris, J P Bourgoin, S Palacin, D. Esteve, C. Urbina, M. Magoga, C. Joachim:'*Electron transfer through a gold-bisthiolterthiophen-gold junction'in Electronic properties of novel materials-science and technology of molecular nanostructures*' ed. H. Kuzmany, J. Fink, M. Mehring, S. Roth (AIP New york 1999) pp421-428
70. J. M. van Ruitenbeek, A. Alvarez, I. Pineyro et al.: Rev. Sci. Instrum. **67**, 108 (1995)
71. H. Grabert, M.H. Devoret: *Single Charge Tunneling,* NATO ASI Series B **294** (Plenum Press New York 1992)
72. J. M. Tour et al: J. Am. Chem. Soc. **117**, 9529 (1995)
73. J. M. Tour: Chem. Rev. **96**, 537 (1996)
74. Y. Q. Xue, S. Datta, S. Hong, et al.: Phys. Rev. B **59**, R7852 (1999)
75. W. Tian, S. Datta, S. Hong, R. Reifenberger, J. I. Henderson, C. P. Kubiak: J Chem. Phys **109**, 2874 (1998)
76. M. Di Ventra, S. T. Pantelides, and N. D. Lang: Phys. Rev. Lett. **84**, 979 (2000)
77. M. Di Ventra, S. T. Pantelides, and N. D. Lang: Appl. Phys. Lett. **76**, 3448 (2000)
78. S. N. Yaliraki, M. Kemp, M. A. Ratner: J. Am. Chem. Soc. **121**, 3428 (1999)
79. S. N. Yaliraki, A. E. Roitberg, C. Gonzalez, V. Mujica, M. A. Ratner: J. Chem. Phys. **111**, 6997 (1999).
80. S. N. Yaliraki and M. A. Ratner: J. Chem. Phys. **109**, 5036 (1998)
81. R. Wiesendanger, H.J. Guentherodt: *Scanning Tunneling Microscopy III*(Springer Verlag,Berlin 1993)
82. H. Ness, A. J. Fisher: Phys. Rev. B **56**, 12469 (1997)
83. C. Dekker, S. J. Tans, B. Oberndorff, R. Meyer, L. C. Venema: Synth. Met. **84**, 853 (1997)
84. C. Joachim, J. F. Vinuesa: Europhys. Lett. **33**, 635 (1996)
85. M. Magoga, C. Joachim: Phys. Rev. B **56**, 4722 (1997)
86. J.P. Bourgoin, C. Kergueris, G. Deniau: unpublished results
87. V. Mujica, M. Kemp, M. A. Ratner: J. Chem. Phys. **101**, 6849 (1994);V. Mujica, M. Kemp, A. Roitberg, and M. Ratner: ibid.**104**, 7296 (1996)
88. E.G. Emberly, G. Kirczenow: Phys. Rev. B **58**, 10911(1998)
89. S. Datta: *Electronic Transport in Mesoscopic Systems* (Cambridge University Press, Cambridge, 1995)
90. D. Porath, Y. Levi, M. Tarabiah, O. Millo: Phys. Rev. B **56**, 1 (1997)
91. C. Zhou, M. R. Deshpande, M. A. Reed, L. Jones II, J. M. Tour: Appl. Phys. Lett. **71,** 611 (1997)
92. C. Bureau, C. Kergueris, J. P. Bourgoin: unpublished results
93. M. Magoga: '*étude du transport tunnel dans une seule molécule',* thèse de l'Université Paul Sabatier, Toulouse 1999
94. H. Ness, A. J. Fisher: Appl. Surf. Sci. **162**, 613 (2000)
95. H. Ness, A. J. Fisher: Phys. Rev. Lett. **83**, 452 (1999)
96. E. G. Emberly, G. Kirczenow: Phys. Rev. B **61**, 5740 (2000)
97. Z.G. Yu, D.L. Smith, A. Saxena, A.R. Bishop: Phys. Rev. B **59**, 1601 (1999)
98. T. Vondrak, C. J. Cramer, X. Y. Zhu: J. Phys. Chem. B 103, 8915 (1999)

99. A. Nitzan, J. Jortner, J. Wilkie, et al.: J. Phys. Chem. B 104, 5661 (2000)
100. D. Segal, A. Nitzan, W.B. Davis et al.: J. Phys. Chem. B **104** 3817 (2000)
101. C. Joachim, C. Bergaud, H. Pinna, et al.: Annals of the New York Academy of Sciences **852**, 243 (1998)
102. C. Joachim, J. K. Gimzewski: Chem. Phys. Lett. **265,** 353 (1997)
103. J. K. Gimzewski, C. Joachim: Science **283**, 1683 (1999)
104. E. G. Emberly, G. Kirczenow: J. Appl. Phys.**88**, 5281(2000)

Luttinger Liquid Behavior
in Metallic Carbon Nanotubes

R. Egger[1], A. Bachtold[2], M.S. Fuhrer[2], M. Bockrath[3], D.H. Cobden[4]
, and P.L. McEuen[2]

[1] Fakultät für Physik, Universität Freiburg, D-79104 Freiburg, Germany
[2] Department of Physics, University of California, Berkeley, CA 94720, USA
[3] Department of Physics, Harvard University, Cambridge, MA 02138, USA
[4] Department of Physics, Warwick University, Coventry, CV4 7AL, UK

Abstract. Coulomb interaction effects have pronounced consequences in carbon nanotubes due to their 1D nature. In particular, correlations imply the breakdown of Fermi liquid theory and typically lead to Luttinger liquid behavior characterized by pronounced power-law suppression of the transport current and the density of states, and spin-charge separation. This paper provides a review of the current understanding of non-Fermi liquid effects in metallic single-wall nanotubes (SWNTs). We provide a self-contained theoretical discussion of electron-electron interaction effects and show that the tunneling density of states exhibits power-law behavior. The power-law exponent depends on the interaction strength parameter g and on the geometry of the setup. We then show that these features are observed experimentally by measuring the tunneling conductance of SWNTs as a function of temperature and voltage. These tunneling experiments are obtained by contacting metallic SWNTs to two nanofabricated gold electrodes. Electrostatic force microscopy (EFM) measurements show that the measured resistance is due to the contact resistance from the transport barrier formed at the electrode/nanotube junction. These EFM measurements show also the ballistic nature of transport in these SWNTs. While charge transport can be nicely attributed to Luttinger liquid behavior, spin-charge separation has not been observed so far. We briefly describe a transport experiment that could provide direct evidence for spin-charge separation.

1   Introduction

The electronic properties of one-dimensional (1D) metals have attracted considerable attention for fifty years by now. Starting with the work of Tomonaga in 1950 [1] and later by Luttinger [2], it has become clear that the electron-electron interaction destroys the sharp Fermi surface and leads to a breakdown of the ubiquituous Fermi liquid theory pioneered by Landau [3]. This breakdown is signalled by a vanishing quasiparticle weight Z_F in the presence of arbitrarily weak interactions. The resulting non-Fermi liquid state is commonly called Luttinger liquid (LL), or sometimes Tomonaga-Luttinger liquid. The name "Luttinger liquid" was coined by Haldane [4] to describe the universal low-energy properties of one-dimensional conductors. Universality means that the physical properties do not depend on details of the model, the interaction potential, etc., but instead are only characterized by a few parameters (critical exponents). The range of validity of the LL model is usually set by $E \ll D$, where D is an electronic bandwidth parameter and E is the relevant energy scale, namely either the thermal scale $k_B T$ or the applied voltage eV. Quite remarkably, the LL concept is believed to hold for arbitrary statistical properties of the particles, e.g. both for fermions and bosons. It provides a paradigm for non-Fermi liquid physics and may have some relevance also for higher-dimensional systems, e.g. in relation to high-temperature superconductivity.

In the model studied by Tomonaga and Luttinger, a special dispersion relation for the noninteracting problem was assumed, where one linearizes around the two Fermi vectors $\pm k_F$ present in 1D. At sufficiently low energy scales, such a procedure should clearly be possible. In fact, we will see below that in a nanotube the dispersion relation is highly linear anyways. Assuming a linear dispersion relation composed of left- and right-moving particles with Fermi velocity v_F, one can equivalently express the noninteracting problem in terms of collective plasmon (density wave) excitations. Technically, in the "bosonization" language [5], for the simplest case of a spinless single-channel system, these bosonic excitations can be expressed in terms of a displacement field $\theta(x)$ such that the density fluctuations are $\rho(x) = \pi^{-1/2} \partial_x \theta(x)$. Electron-electron interactions then describe a bilinear coupling of these density fluctuations, and therefore the full interacting problem can be written as a free theory in the displacement field:

$$H = \frac{\hbar v_F}{2} \int dx \left(\Pi^2(x) + \frac{1}{g^2} [\partial_x \theta(x)]^2 \right) , \tag{1}$$

where $\Pi(x)$ is the canonical momentum to the field $\theta(x)$. In the long-wavelength limit, one can approximate the Fourier transform $\widetilde{V}(k)$ of the 1D interaction potential by a constant $V_0 = \widetilde{V}(0) - \widetilde{V}(2k_F)$, and the dimensionless g parameter in Eq. (1) is given by

$$g = [1 + V_0/\pi \hbar v_F]^{-1/2} . \tag{2}$$

Note that for repulsive interactions we always have $g < 1$, with small g meaning strong interactions. The limit $g = 1$ describes the Fermi gas (not a Fermi liquid), and the limit $g \to 0$ leads to a classical Wigner crystal. The model (1) is equivalent to a set of harmonic oscillators and can therefore be solved exactly. The physical interpretation can be elucidated by the use of the bosonization formula for the electron operator itself [5]. Thereby, the creation operator for a right- or left-moving electron ($r = R/L = \pm$) can be written in the form

$$\psi_r(x) \simeq \frac{1}{\sqrt{2\pi a}} \exp\left(irk_F x + ir\sqrt{\pi}\theta(x) + i\sqrt{\pi} \int^x dx' \Pi(x') \right) , \tag{3}$$

where $a \approx 1/k_F$ is a lattice constant. Using this expression, it is a simple matter to show that the sharp $T = 0$ Fermi surface is smeared out for $g < 1$, with interaction-dependent power laws close to k_F. Physically, this is because the electron is an unstable particle and spontaneously decays into collective plasmon modes. Including the spin-1/2 degree of freedom, one finds that the spin and charge plasmons also decouple and moreover propagate with different velocities $v_c \neq v_s$. This phenomenon is called *spin-charge separation* and implies that the spin and charge degrees of freedom of an electron brought into a LL will spatially separate. Note that in a Fermi liquid $v_c = v_s$ and therefore this characteristic feature will not show up. Spin-charge separation is intrinsically a dynamical phenomenon outside the scope of thermodynamics.

An interesting and closely related issue concerns the fractionalized stable excitations of the LL. While it is easy to establish the spin-charge separation phenomenon in the bosonic plasmon basis, the nature of the expected fundamental "quasiparticles" with fractional statistics, similar to the famous Laughlin quasiparticles in the fractional quantum Hall (FQH) effect, is less clear. In a 1D Hubbard chain, which is known to be a realization of the LL at low temperatures, well-defined spinon and holon excitations exist. For a spinless system, one can establish that quasiparticles scattered by a weak impurity potential have fractional charge ge and a statistical angle πg [6]. Remarkably, the fractional charge can have any – even irrational – value. Furthermore, for the topology of a LL on a ring, a complete characterization of the universal LL theory in terms of fractional-statistics quasiparticles has been provided recently [7].

In view of this discussion, it is understandable that, for many decades, experimentalists have attempted to find LL behavior. In the 1970s, the key interest was focused on quasi-1D organic chain compounds [8], where LL behavior is hard to establish because of complicated 1D-3D crossover phenomena and additional phase transitions into other states. The interest was revived a few years ago, when experimental observations of LL behavior for transport in semiconductor quantum wires [9,10] and for edge states in FQH bars [11,12] were reported. Shortly after the theoretical prediction of LL behavior in metallic carbon nanotubes [13,14], the to-date perhaps cleanest experimental observations of LL behavior were established in transport experiments for single-wall nanotubes (SWNTs) [15,16]. The theory along with the experiments of Ref. [15] will be presented below. By now, there are also several other theoretical proposals for probing the LL state in *bulk* systems, e.g. by investigating the tunneling density of states (TDOS) of a 3D metal in an ultra-strong magnetic field [17], or by studying 2D arrays of regularly stacked nanotubes [18].

Carbon nanotubes were discovered in 1991 by Iijima [19] and have enjoyed exponentially increasing interest since then. The current status of the field has been summarized in a recent Physics World issue [20], see also Ref. [21]. Ignoring the end structure, one may think of a SWNT as a graphene sheet, i.e. a 2D honeycomb lattice made up of C atoms, that is wrapped onto a cylinder, with typical radius of order 1-2 nm and length of several microns. Depending on the helicity of the wrapping, the resulting SWNT is either semiconducting or metallic. In our experimental setup discussed in Sec. 3, these two behaviors can be distinguished as follows. When the conductance G of the tube is measured as a function of a gate voltage V_g, G is virtually independent of V_g for metal tubes, while G varies exponentially with V_g for semiconducting tubes. The discussion in this paper is limited to transport through *metallic* SWNTs, where LL behavior can be expected.

From the special band structure of a graphene sheet [20], one arrives at the characteristic dispersion relation of a metallic SWNT shown in Figure 1. This band structure exhibits two Fermi points $\alpha = \pm$ with a right- and a left-moving ($r = R/L = \pm$) branch around each Fermi point. These branches are highly linear with Fermi velocity $v_F \approx 8 \times 10^5$ m/s. The R- and L-movers arise as linear combinations of the $p = \pm$ sublattice states reflecting the two C atoms in the basis of the honeycomb lattice. The dispersion relation depicted in Fig. 1 holds for energy scales $E < D$, with the bandwidth cutoff scale $D \approx \hbar v_F/R$ for tube radius R. For typical SWNTs, D will be of the order 1 eV. The large overall energy scale together with the structural stability of SWNTs explain their unique potential for revealing LL physics. In contrast to conventional systems, e.g. semiconductor quantum wires, LL effects in SWNTs are not restricted to the meV range but may even be seen at room temperature. An additional advantage is that the approximation introduced by linearizing the dispersion relation in conventional 1D systems is here provided by nature in an essentially exact way. A basic prerequisite of the theory [13] is the *ballistic* nature of transport in SWNTs. Ballistic transport in SWNTs can be unambiguously established by various experiments, see below and Ref. [22–24]. Theoretical analysis [25] has also suggested the absence of a diffusive phase in SWNTs, with the possibility of ballistic transport over distances of several μm.

Besides SWNTs, LL effects have also been observed in the TDOS of multi-wall nanotubes (MWNTs) [26,27]. MWNTs are composed of several concentrically arranged graphene shells, and under the assumption of ballistic transport, the only incomplete screening does not spoil the LL behavior [28]. On the other hand, transport in MWNTs has typical signatures of *diffusive* transport [24,26], and the theoretical situation must be regarded as unsettled at the moment. We shall therefore only discuss (metallic) SWNTs in this review.

The structure of the article is as follows. In Sec. 2, the theoretical description of a metallic SWNT in the ballistic limit is reviewed, where we focus on the low-energy regime $E < D$. We shall derive the scaling forms of the nonlinear dI/dV characteristics for bulk or end tunneling into a nanotube, and point to various experimental setups that can detect correlation effects in the transport. We shall also briefly outline a recent suggestion for a spin-transport experiment that could allow for the experimental verification of spin-charge separation. In Sec. 3, the experimental evidence for LL behavior found so far is reviewed. Finally, in Sec. 4, we summarize and discuss some of the open problems that we are aware of.

2   Luttinger-Liquid Theory for Nanotubes

2.1   Low-Energy Theory: General Approach

The remarkable electronic properties of carbon nanotubes are due to the special bandstructure of the π electrons in graphene. There are only two linearly independent Fermi points αK with $\alpha = \pm$ instead of a continuous Fermi surface. Up

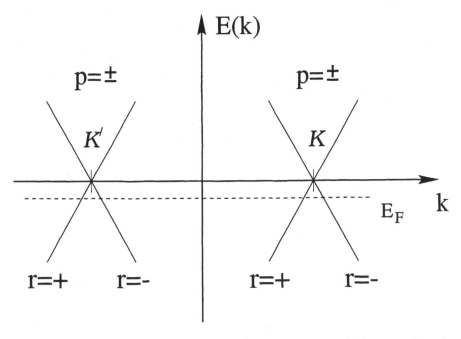

Fig. 1. Schematic bandstructure of a metallic SWNT. A right- and left-moving branch ($r = \pm$) is found near each of the two Fermi points $k = \alpha k_F$ with $\alpha = \pm$, corresponding to K and K', respectively. Right- and left-movers arise as linear combinations of the sublattices $p = \pm$. The Fermi energy (dashed line) is shifted away from neutrality by doping and/or external gates.

to energy scales $E < D \approx 1$ eV, the dispersion relation around the Fermi points is, to a very good approximation, linear. Since the basis of the honeycomb lattice contains two atoms, there are two sublattices $p = \pm$, and hence two degenerate Bloch states

$$\varphi_{p\alpha}(\boldsymbol{r}) = (2\pi R)^{-1/2} \exp(-i\alpha \boldsymbol{K}\boldsymbol{r}) \tag{4}$$

at each Fermi point $\alpha = \pm$. Here $\boldsymbol{r} = (x, y)$ lives on the sublattice p under consideration, and we have already anticipated the correct normalization for nanotubes. The Bloch functions are defined separately on each sublattice such that they vanish on the other. One can then expand the electron operator in terms of these Bloch functions. The resulting effective low-energy theory of graphene is the 2D massless Dirac hamiltonian. This result can also be derived in terms of $\boldsymbol{k} \cdot \boldsymbol{p}$ theory.

Wrapping the graphene sheet onto a cylinder then leads to the generic bandstructure of a metallic SWNT shown in Fig. 1. Writing the Fermi vector as $\boldsymbol{K} = (k_F, p_F)$, where the x-axis is taken along the tube direction and the circumferential variable is $0 < y < 2\pi R$, quantization of transverse motion now allows for a contribution $\propto \exp(imy/R)$ to the wavefunction. However, excitation of angular momentum states other than $m = 0$ costs a huge energy of order $D \approx 1$ eV. In an effective low-energy theory, we may thus omit all transport bands except $m = 0$ (assuming that the SWNT is not excessively doped). Evidently, the nanotube forms a 1D quantum wire with only two transport bands intersecting the Fermi energy. This strict one-dimensionality is fulfilled up to remarkably high energy scales (eV) here, in contrast to conventional 1D conductors. The electron operator for spin $\sigma = \pm$ is then written as

$$\Psi_\sigma(x,y) = \sum_{p\alpha} \varphi_{p\alpha}(x,y)\,\psi_{p\alpha\sigma}(x)\,, \tag{5}$$

which introduces slowly varying 1D fermion operators $\psi_{p\alpha\sigma}(x)$ that depend only on the x coordinate. Neglecting Coulomb interactions for the moment, the hamiltonian is:

$$H_0 = -\hbar v_F \sum_{p\alpha\sigma} p \int dx\, \psi^\dagger_{p\alpha\sigma}\partial_x\psi_{-p\alpha\sigma}\,. \tag{6}$$

Switching from the sublattice ($p = \pm$) description to the right- and left-movers ($r = \pm$) indicated in Fig. 1 implies two copies ($\alpha = \pm$) of massless 1D Dirac hamiltonians for each spin direction. Therefore a perfectly contacted and clean SWNT is expected to have the quantized conductance $G_0 = 4e^2/h$. Due to the difficulty of fabricating sufficiently good contacts, however, this value has not been experimentally observed so far. (We note that the conductance quantum $2e^2/h$ seen in recent MWNT experiments by Frank et al. [29] is anomalous and does not correspond to the expected value of G_0.) Remarkably, other spatial oscillation periods than the standard wavelength $\lambda = \pi/k_F$ are possible. From Fig. 1 we observe that the wavelengths

$$\lambda = \pi/k_F, \quad \pi/|q_F|, \quad \pi/(k_F \pm q_F) \tag{7}$$

could occur, where the doping determines the wavevector $q_F \equiv E_F/\hbar v_F$. Which of the wavelengths (7) is ultimately realized sensitively depends on the interaction strength [14].

2.2   Electron-Electron Interactions

Let us now examine Coulomb interactions mediated by an arbitrary potential $U(r - r')$. The detailed form of this potential will depend on properties of the substrate, nearby metallic gates, and the geometry of the setup. In the simplest case, bound electrons and the effects of an insulating substrate are described by a dielectric constant κ, and for an externally unscreened Coulomb interaction,

$$U(r - r') = \frac{e^2/\kappa}{\sqrt{(x - x')^2 + 4R^2 \sin^2[(y - y')/2R] + a_z^2}}\,, \tag{8}$$

where $a_z \approx a$ denotes the average distance between a $2p_z$ electron and the nucleus, i.e. the "thickness" of the graphene sheet. We neglect relativistic effects

like retardation or spin-orbit coupling in the following. Electron-electron interactions are then described by the second-quantized hamiltonian

$$H_I = \frac{1}{2} \sum_{\sigma\sigma'} \int d\mathbf{r} \int d\mathbf{r}' \, \Psi_\sigma^\dagger(\mathbf{r})\Psi_{\sigma'}^\dagger(\mathbf{r}')U(\mathbf{r}-\mathbf{r}')\Psi_{\sigma'}(\mathbf{r}')\Psi_\sigma(\mathbf{r}) \,. \tag{9}$$

The interaction (9) can be reduced to a 1D form by inserting the expansion (5) for
the electron field operator. The reason to do so is the large arsenal of theoretical
methods readily available for 1D models. The result looks quite complicated at
first sight:

$$H_I = \frac{1}{2} \sum_{pp'\sigma\sigma'} \sum_{\{\alpha_i\}} \int dx dx' \, V^{pp'}_{\{\alpha_i\}}(x-x')\psi^\dagger_{p\alpha_1\sigma}(x)\psi^\dagger_{p'\alpha_2\sigma'}(x')\psi_{p'\alpha_3\sigma'}(x')\psi_{p\alpha_4\sigma}(x) \,,$$
$$\tag{10}$$

with the 1D interaction potentials

$$V^{pp'}_{\{\alpha_i\}}(x-x') = \int dy dy' \, \varphi^*_{p\alpha_1}(\mathbf{r})\varphi^*_{p'\alpha_2}(\mathbf{r}')U(\mathbf{r}-\mathbf{r}'+pd\delta_{p,-p'})\varphi_{p'\alpha_3}(\mathbf{r}')\varphi_{p\alpha_4}(\mathbf{r}) \,. \tag{11}$$

These potentials only depend on $x-x'$ and on the 1D fermion quantum numbers.
For interactions involving different sublattices $p \neq p'$ for $\mathbf{r}$ and $\mathbf{r}'$ in Eq. (9), one
needs to take into account the shift vector $\mathbf{d}$ between sublattices.

To simplify the resulting 1D interaction (10), we now exploit momentum
conservation, assuming $E_F \neq 0$ so that Umklapp electron-electron scattering
can be ignored. We then have "forward scattering" processes, where $\alpha_1 = \alpha_4$
and $\alpha_2 = \alpha_3$. In addition, "backscattering" processes may be important, where
$\alpha_1 = -\alpha_2 = \alpha_3 = -\alpha_4$. We first define the potential

$$V_0(x-x') = \int_0^{2\pi R} \frac{dy}{2\pi R} \int_0^{2\pi R} \frac{dy'}{2\pi R} \, U(\mathbf{r}-\mathbf{r}') \,. \tag{12}$$

For the unscreened Coulomb interaction (8), this can be explicitly evaluated
[14]. From Eqs. (11) and (4), the forward scattering interaction potential reads
$V_0(x) + \delta_{p,-p'}\delta V_p(x)$, with

$$\delta V_p(x) = \int_0^{2\pi R} \frac{dy dy'}{(2\pi R)^2} [U(x+pd_x, y-y'+pd_y) - U(x, y-y')] \,, \tag{13}$$

which is only present if $\mathbf{r}$ and $\mathbf{r}'$ are located on different sublattices. Thereby important information about the discrete nature of the graphite network is retained
despite the low-energy continuum approximation. Since $V_0(x)$ treats both sublattices on equal footing, the resulting part of the forward scattering interactions
couples only the total 1D electron densities,

$$H_I^{(0)} = \frac{1}{2} \int dx dx' \, \rho(x)V_0(x-x')\rho(x') \,, \tag{14}$$

where the 1D density is $\rho = \sum_{p\alpha\sigma} \psi^\dagger_{p\alpha\sigma}\psi_{p\alpha\sigma}$. This part of the electron-electron interaction is the most important one and will be seen to imply LL behavior. Note that it is entirely due to the *long-ranged* tail of the Coulomb interaction. All the remaining residual interactions come from short-ranged interaction processes, and since these are intrinsically averaged over the circumference of the tube, their amplitude is quite small and will (at worst) only cause exponentially small gaps. A related general discussion can be found in Ref. [30].

For $|x| \gg a$, detailed analysis shows that $\delta V_p(x) = 0$. However, for $|x| \leq a$, an additional term beyond Eq. (14) arises due to the hard core of the Coulomb interaction. At such small length scales, the difference between inter- and intra-sublattice interactions matters. To study this term, one should evaluate $\delta V_p(0)$ from microscopic considerations. One then finds the additional forward scattering contribution [14]

$$H_I^{(1)} = -f \int dx \sum_{p\alpha\alpha'\sigma\sigma'} \psi^\dagger_{p\alpha\sigma}\psi^\dagger_{-p\alpha'\sigma'}\psi_{-p\alpha'\sigma'}\psi_{p\alpha\sigma} , \tag{15}$$

where $f/a = \gamma_f e^2/R$. An estimate for armchair SWNTs yields $\gamma_f \approx 0.05$. Since these short-ranged interaction processes are averaged over the circumference of the tube, $f \propto 1/R$, and hence f is very small. A similar reasoning applies to the backscattering contributions $\alpha_1 = -\alpha_2 = \alpha_3 = -\alpha_4$ in Eq. (10). Because of a rapidly oscillating phase factor, the only non-vanishing contribution comes again from $|x - x'| \leq a$, and we can effectively take a local interaction. Furthermore, only the part of the interaction which does not distinguish among the s⎯blattices is relevant and leads to

$$H_I^{(2)} = b \int dx \sum_{pp'\alpha\sigma\sigma'} \psi^\dagger_{p\alpha\sigma}\psi^\dagger_{p'-\alpha\sigma'}\psi_{p'\alpha\sigma'}\psi_{p-\alpha\sigma} . \tag{16}$$

For the unscreened interaction (8), $b/a = \gamma_b e^2/R$ with $\gamma_b \approx \gamma_f$. For externally screened Coulomb interaction, one may have $b \gg f$.

Progress can then be made by employing the *bosonization* approach [5]. For that purpose, one first needs to bring the non-interacting hamiltonian (6) into the standard form of the 1D Dirac model. This is accomplished by switching to right- and left-movers ($r = \pm$) which are linear combinations of the sublattice states $p = \pm$. In this representation, a bosonization formula generalizing Eq. (3) applies, now with four bosonic phase fields $\theta_a(x)$ and their canonical momenta $\Pi_a(x)$. The four channels are obtained from combining charge and spin degrees of freedom as well as symmetric and antisymmetric linear combinations of the two Fermi points, $a = c+, c-, s+, s-$. The bosonized expressions for H_0 and $H_I^{(0)}$ read

$$H_0 = \sum_a \frac{\hbar v_F}{2} \int dx \left[\Pi_a^2 + g_a^{-2}(\partial_x\theta_a)^2 \right] \tag{17}$$

$$H_I^{(0)} = \frac{2}{\pi} \int dx dx' \, \partial_x\theta_{c+}(x)V_0(x - x')\partial_{x'}\theta_{c+}(x') . \tag{18}$$

The bosonized form of $H_I^{(1,2)}$ [13] leads to nonlinearities in the θ_a fields for $a \neq c+$. Although bosonization of Eq. (6) gives $g_a = 1$ in Eq. (17) [see also Eq. (1)], interactions will renormalize these parameters. In particular, in the long-wavelength limit, $H_I^{(0)}$ can be incorporated into H_0 by putting

$$g_{c+} \equiv g = \left\{ 1 + 4\widetilde{V}_0(k \simeq 0)/\pi\hbar v_F \right\}^{-1/2} \leq 1 , \tag{19}$$

while for all other channels, the coupling constant f gives rise to the tiny renormalization $g_{a \neq c+} = 1 + f/\pi\hbar v_F \simeq 1$. The plasmon velocities of the four modes are $v_a = v_F/g_a$, and hence the charged ($c+$) mode propagates with significantly higher velocity than the three neutral modes.

For the long-ranged interaction (8), the logarithmic singularity in $\widetilde{V}_0(k)$ requires the infrared cutoff $k = 2\pi/L$ due to the finite length L of the SWNT, resulting in:

$$g = \left\{ 1 + \frac{8e^2}{\pi\kappa\hbar v_F} \ln(L/2\pi R) \right\}^{-1/2} . \tag{20}$$

Since $\hbar c/e^2 \simeq 137$, we get with $v_F = 8 \times 10^5$ m/s the estimate $e^2/\hbar v = (e^2/\hbar c)(c/v) \approx 2.7$, and therefore g is typically in the range 0.2 to 0.3. This estimate does only logarithmically depend on L and R, and should then apply to basically all SWNTs studied at the moment (where $L/R \approx 10^3$). The LL parameter g predicted by Eq. (20) can alternatively be written in the form

$$g = \left(1 + \frac{2E_c}{\Delta} \right)^{-\frac{1}{2}} , \tag{21}$$

where E_c is the charging energy and Δ the single-particle level spacing. For our experimental setup described in Sec. 3, the theoretically expected LL parameter is then estimated as $g_{\mathrm{th}} \approx 0.28$. The very small value of g obtained here implies that an individual metallic SWNT on an insulating substrate is a strongly correlated system displaying very pronounced non-Fermi liquid effects.

It is clear from Eqs. (17) and (18) that for $f = b = 0$, a SWNT constitutes a realization of the LL. We therefore have to address the effect of the nonlinear terms associated with the coupling constants f and b. This can be done by means of the renormalization group approach. Together with a solution via Majorana refermionization, this route allows for the complete characterization of the non-Fermi-liquid ground state of a clean nanotube [14]. From this analysis, we find that for temperatures above the exponentially small energy gap

$$k_B T_b = D \exp[-\pi\hbar v_F/\sqrt{2}b] \tag{22}$$

induced by electron-electron backscattering processes, the SWNT is adequately described by the LL model, and $H_I^{(1,2)}$ can effectively be neglected. A rough order-of-magnitude estimate is $T_b \approx 0.1$ mK. In the remainder, we focus on temperatures well above T_b.

2.3   Bulk and End Tunneling: Scaling Functions and Exponents

Under typical experimental conditions, the contact between a SWNT and the attached (Fermi-liquid) leads is not perfect and the conductance is limited by electron tunneling into the SWNT, which in turn is governed by the TDOS. The TDOS exhibits power-law behavior and is strongly suppressed at low energy scales. The power-law exponent $\alpha > 0$ depends on the geometry of the particular experiment: If one tunnels into the end of a SWNT, the exponent α_{end} is generally larger than the bulk exponent α_{bulk}, since electrons can move in only one direction to accomodate the incoming additional electron. The end-tunneling exponent can be easily obtained from the open boundary bosonization technique [5]. It follows that close to the boundary (taken at $x = 0$), i.e. for $\max(x, x') \ll v_F t$, the single-electron Greens function is of the form

$$\langle \Psi(x, t)\Psi^\dagger(x', 0)\rangle \propto t^{-(1/g+3)/4} . \tag{23}$$

The boundary scaling dimension of the electron field operator is therefore $\bar{\Delta} = \frac{1}{8g} + \frac{3}{8}$, as opposed to its bulk scaling dimension $\Delta = \frac{1}{16}\left(\frac{1}{g} + g\right) + \frac{3}{8}$. Making use of the text-book definition of the TDOS as the imaginary part of the electron Greens function, we find from Eq. (23) that the TDOS indeed vanishes as a power law with energy,

$$\rho(E) \propto (E/D)^\alpha , \tag{24}$$

where the exponent α is given by the end-tunneling exponent

$$\alpha_{\text{end}} = 2\bar{\Delta} - 1 = \left(\frac{1}{g} - 1\right)/4 . \tag{25}$$

Similarly one may derive the bulk-tunneling exponent:

$$\alpha_{\text{bulk}} = 2\Delta - 1 = \left(\frac{1}{g} + g - 2\right)/8 . \tag{26}$$

Since $\alpha > 0$ for $g < 1$, the TDOS vanishes as the energy scale E approaches zero in both cases. For a Fermi liquid, however, both exponents are zero.

If transport is limited by tunneling through a weak contact from a metal electrode to the SWNT, the full nonlinear and temperature-dependent differential conductance $G(V, T) = dI/dV$ can be evaluated in closed form. If V denotes the voltage drop across the weak link, one obtains

$$G(V, T) = AT^\alpha \cosh\left(\frac{eV}{2k_BT}\right)\left|\Gamma\left(\frac{1+\alpha}{2} + \frac{ieV}{2\pi k_BT}\right)\right|^2 , \tag{27}$$

where Γ denotes the gamma function and A is a nonuniversal prefactor depending on details of the junction. The exponent α is either the end- or the bulk-tunneling exponent depending on the experimental geometry. If the leads are at finite temperature, the conductance is given by a convolution of Eq. (27) and the derivative of the Fermi function:

$$-df/dE = \frac{1}{4k_B T \cosh^2(eV/2k_B T)} .$$

Remarkably, the quantity $T^{-\alpha}G(V,T)$ should then be a *universal* scaling function of the variable $eV/k_B T$ alone. This scaling is seen experimentally as discussed in Sec. 3.

2.4   Crossed Nanotubes

More spectacular correlation effects can be observed in more complicated geometries. The simplest example is provided by *crossed nanotubes* [31] which have recently been studied experimentally [27,32]. The geometry is shown in Figure 2, where we assume a pointlike contact of two clean metallic SWNTs characterized by the same g parameter. External reservoirs can be incorporated by imposing Sommerfeld-like radiative boundary conditions [33] close to the contacts (for simplicity, we sketch the theory for the spinless single-channel case). This approach offers a general and powerful route to studying multi-terminal Landauer-Büttiker geometries for correlated 1D systems. Applying the two-terminal voltage V_i along conductor $i = 1, 2$, the boundary conditions read

$$\left(\frac{1}{g^2}\partial_x \pm \frac{1}{v_F}\partial_t \right) \langle \theta_i(x = \mp L/2, t) \rangle = \frac{eV_i}{\sqrt{\pi}\hbar v_F} . \tag{28}$$

These boundary conditions fix the average densities of injected particles. Outgoing particles are assumed to enter the reservoirs without reflection.

Let us now consider a point-like coupling at, say, $x = 0$. Such a contact causes (at least) two different coupling mechanisms. First, there arises an *electrostatic interaction* $H_c^{(1)} \propto \rho_1(0)\rho_2(0)$. Bosonization shows that the only important part is

$$H_c^{(1)} = \lambda \cos[\sqrt{4\pi}\,\theta_1(0)] \cos[\sqrt{4\pi}\,\theta_2(0)] , \tag{29}$$

which becomes relevant for sufficiently strong interactions, $g < 1/2$. The second potentially important process is *single-electron tunneling* from one conductor into the other. Notably, tunneling is always irrelevant for $g < 1$, and (unless the contact is very good) can therefore be treated in perturbation theory. In other words, tunneling is expected to have only a very minor effect here, and we shall hence focus on the effect of $H_c^{(1)}$ specified in Eq. (29). Again, for $g > 1/2$, this term can also be treated perturbatively, but for the interesting strong-interaction case $g < 1/2$, qualitatively new features in the transport emerge.

To investigate this situation further, we switch to the linear combinations $\theta_\pm(x) = [\theta_1(x) \pm \theta_2(x)]/\sqrt{2}$, whence the hamiltonian decouples into the sum $H_+ + H_-$ with

$$H_\pm = \frac{\hbar v_F}{2} \int dx \left\{ \Pi_\pm^2 + \frac{1}{g^2}(\partial_x \theta_\pm)^2 \right\} \pm (\lambda/2) \cos\left[\sqrt{8\pi}\,\theta_\pm(0) \right] . \tag{30}$$

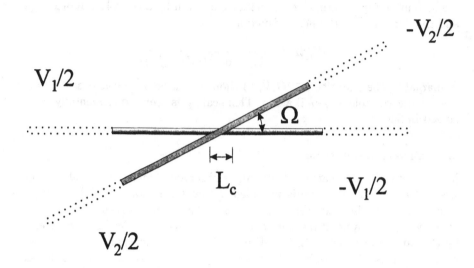

Fig. 2. Crossed nanotube setup. By variation of the angle Ω, the contact length L_c can be changed. We consider a pointlike contact, $L_c \leq a$.

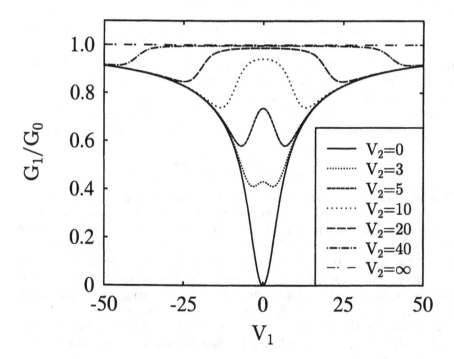

Fig. 3. Conductance $G_1/G_0 \equiv I_1/(e^2 V_1/h)$ for $g = 1/4, T = 0$, and several values of the cross voltage V_2. The overall energy scale is set by the coupling λ.

Effective boundary conditions (28) for the fields $\theta_\pm$ are found by simply replacing $V_{1,2} \to (V_1 \pm V_2)/\sqrt{2}$. Therefore we are left with two completely decoupled systems, each of which is formally identical to the problem of an elastic potential scatterer embedded into a spinless LL with effectively doubled interaction strength parameter $g' = 2g$. The hamiltonian (30) has been discussed previously by Kane and Fisher [34], and the exact solution under the boundary condition (28) has recently been given by boundary conformal field theory methods [35]. This solution applies for arbitrary g, T, V, and λ.

The conductance $G_1 = I_1/(e^2 V_1/h)$ for $g = 1/4$ at zero temperature is plotted as a function of V_1 and V_2 in Fig. 3. Contrary to what is found in the uncorrelated case, G_1 is extremely sensitive to both V_1 and V_2 (in a Fermi liquid, G_1 is simply constant). For $V_2 = 0$, transport becomes fully suppressed for $V_1 \to 0$, with a g-dependent perfect *zero-bias* anomaly (ZBA). Remarkably, there is a suppression of the current if $|V_1| = |V_2|$, which is observed as a "dip" in $G_1(V_1)$ for fixed V_2. This effect can be rationalized in terms of a partial dynamical pinning of charge density waves in tube 1 due to commensurate charge density waves in tube 2. The consequence is that the ZBA dip at $V_1 = 0$ is turned into a peak by increasing the cross voltage V_2. The pronounced and nonlinear sensitivity of $G_1(V_1, V_2)$ to V_2 is a distinct fingerprint for LL behavior. Qualitatively, all these features have been observed in a very recent experiment by Kim et al. [27] on crossed MWNTs.

2.5 Spin Transport

The ultimate hallmark of a LL is electron fractionalization and spin-charge separation. So far no unambiguous experimental verification of spin-charge separation in a LL has been published, and carbon nanotubes might offer the possibility to do so. The standard approach via photoemission is clearly not suitable here since one should work on a single SWNT. Alternatively, a spin transport experiment will be described below that should reveal spin-charge separation in a clear manner [36]. In such an experiment, one needs to measure the $I - V$ characteristics of a SWNT in weak contact to two *ferromagnetic* reservoirs, where the angle ϕ between the ferromagnet magnetization directions $\hat{m}_1$ and $\hat{m}_2$, i.e. $\cos \phi = \hat{m}_1 \cdot \hat{m}_2$, can take an arbitrary value $0 \leq \phi \leq \pi$. A corresponding experiment for $\phi = 0, \pi$ has recently been performed for a MWNT [37].

Spin transport has been studied in detail for Fermi liquids. For the proposed geometry of a metal connected to ferromagnetic leads via tunnel junctions, Brataas et al. [38] have computed the ϕ-dependence of the current. Assuming identical junction and ferromagnet parameters, they obtain

$$\frac{I(\phi)}{I(0)} = 1 - P^2 \frac{\tan^2(\phi/2)}{\tan^2(\phi/2) + Y} , \tag{31}$$

where the polarization $0 \leq P \leq 1$ parametrizes the difference in the spin-dependent DOS of a ferromagnetic reservoir, and $Y \geq 1$ is related to the spin-mixing conductance [38]. The result (31) shows that for any $\phi > 0$ the current will be suppressed due to the spin accumulation effect [39]. The maximum suppression, namely by a factor $1 - P^2$, occurs for antiparallel magnetizations, $\phi = \pi$.

If one has spin-charge separation, detailed analysis [36] shows that the current is still properly described by Eq. (31), though with two important differences. First, the current $I(0)$ for parallel magnetizations will carry the usual power-law suppression factor $(V/D)^{\alpha/2}$, where $\alpha > 0$ is the bulk/end tunneling exponent. More importantly, the quantity Y will now be V- and T-dependent, with a divergence as $V, T \to 0$ according to $Y \propto [\max(eV, k_B T)/D]^{-\alpha}$. Therefore the spin accumulation effect, i.e. the suppression of the current by changing ϕ away from zero, will be totally destroyed by spin-charge separation, except for $\phi = \pi$. This qualitative difference to a Fermi liquid should be easily detectable and can serve as a signature of spin-charge separation.

3   Experimental Evidence for Luttinger Liquid

In this section, we show first with electrostatic force microscopy (EFM) that metallic nanotubes are ballistic conductors, an important ingredient for the possible observation of LL behavior. When nanotubes are attached to metallic electrodes, EFM shows that a barrier is formed at the nanotube/metal interface. This fact is then exploited to observe LL behavior in nanotube devices via the TDOS. We show experimentally that the TDOS indeed exhibits power-law behavior in metallic SWNTs. This is observed by mesuring the tunneling conductance of nanotube/metal interfaces as a function of temperature and voltage.

3.1   Electrostatic Force Microscopy of Electronic Transport in Carbon Nanotubes

Samples are fabricated on a backgated substrate consisting of degenerately doped silicon capped with 1 μm SiO_2. SWNTs synthesized via laser ablation are ultrasonically suspended in dichloroethane, and the resulting suspension is placed on the substrate for approximately 15 seconds, then washed off with isopropanol. An array of structure, each consisting of two Cr/Au electrodes, is fabricated using electron beam lithography. Samples that have a measurable resistance between the electrodes are selected with a prober. An AFM is then used to choose samples that have only one nanotube rope between the electrodes. Objects whose height profile is consistent with single SWNTs (1-2 nm) are preferentially selected. An example of a SWNT rope contacted by two electrodes is shown in Fig. 4(a). Since the success of this contacting scheme works by chance, it is obvious that the yield is low. However, since a large array of structures can readily be fabricated, this scheme has turned out to be very convenient.

We continue by reviewing the EFM technique [40] which is used to directly probe the nature of conduction in SWNTs. An AFM tip with a voltage V_{tip} is scanned over a nanotube sample, see Fig. 4(b). The electrostatic force between the tip and the sample is given by

$$F = \frac{1}{2} \frac{dC}{dz} (V_{\text{tip}} + \phi - V_s)^2 , \tag{32}$$

(a)   **(b)**

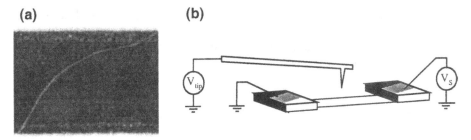

Fig. 4. (a) Topographic AFM image of a 2.5 nm diameter bundle of SWNTs which is seen spanning between two gold electrodes. The separation between the electrodes is 1 μm. (b) Experimental setup for EFM. A conducting AFM cantilever is scanned above the device, which consists of a nanotube contacted by two gold electrodes. Adapted from Ref. [24].

where V_s is the voltage within the sample, ϕ is the work function difference between the tip and sample, and C is the tip-sample capacitance. The tip is held at constant height above the surface by first making a line-scan of the topography of the surface using intermittent-contact AFM, and then making a second pass with the tip held at a fixed distance above the measured topographic features. In order to detect the electrostatic force, the cantilever is made to oscillate by an AC potential that is applied to the sample at the resonant frequency of the cantilever. This produces an AC force on the cantilever proportional to the local AC potential $V_s(w)$ beneath the tip:

$$F_{ac}(w) = \frac{dC}{dz}(V_{\text{tip}} + \phi)\, V_s(w)\,. \tag{33}$$

The resulting oscillation amplitude is recorded using an external lock-in amplifier; the signal is proportional to $V_s(w)$. Calibration of this signal is made by applying a uniform $V_s(w)$ to the whole sample and measuring the response of the cantilever.

EFM yields a signal that is proportional to the local voltage within the nanotube circuit. However, the signal is also proportional to the derivative of the local capacitance. This will vary as the geometry changes, yielding e.g. different signals over a nanotube than over a contact at the same potential. However, dC/dz does not vary appreciably as a function of distance along the nanotube. The measured signal should thus accurately reflect the local voltage within the nanotube.

3.2   Ballistic Transport in Metallic SWNTs

Next we discuss measurements of the device shown in Fig. 4. The resistance of this 2.5 nm diameter bundle is 40 kΩ and has no significant gate voltage dependence. We have also measured the current at large biases – the current saturates at 50 μA. This is in agreement with recent work by Yao, Kane and Dekker [41] where the current was observed to be limited to 25 μA per metallic nanotube due to optical or zone-boundary phonon scattering. We therefore conclude that the current is carried by 2 metallic SWNTs in the bundle.

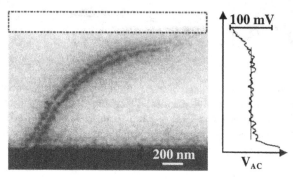

Fig. 5. EFM image of the same bundle of SWNTs shown in Fig. 4. An AC potential of 100 mV is applied to the lower electrode. The upper electrode indicated by the box is grounded. The AC-EFM signal is flat along the length of the SWNT bundle, indicating that the potential drops occur at the contacts, and not along the bundle length. A trace of the potential as a function of vertical position in the image is also shown. Adapted from Ref. [24].

Figure 5 shows the EFM image of this SWNT bundle, as well as a line trace along the backbone of the bundle. The potential is flat over its length, indicating that within our measurement accuracy there is no measurable intrinsic resistance. Taking into account the finite measurement resolution, we estimate that R_i of the bundle is at most 3 kΩ. The contact resistances are measured to be approximately 28 kΩ and 12 kΩ for the upper and lower contacts, respectively. The conductance of the tube has been controlled to no change when the tip scans over it. The original data have a background signal due to stray capacitive coupling of the tip to the large metal electrodes. The image in Fig. 5 is shown with the background signal subtracted according to the procedure described in Ref. [24].

Using the four-terminal Landauer formula, $R = (h/4e^2)(1 - T_i)/T_i$ per nanotube, where T_i is the transmission coefficient for electrons along the length of the nanotube, we find that T_i is larger than 0.5. This indicates that the majority of electrons are transported through the bundle with no scattering. Therefore transport is ballistic at room temperature over a length of $> 1\mu$m. This confirms the theoretical predictions of very weak scattering in metallic SWNTs [25,42]. This is also in agreement with previous low-temperature transport measurements which indicate that long metallic SWNTs may behave as single quantum dots [22,23], and room-temperature measurements of metallic SWNTs which sometimes exhibit low two-terminal resistance [43]. The dominant portion of the overall resistance of 40 kΩ thus comes from the contacts, indicating that the transmission coefficients for entering and leaving the bundle are significantly less than one and the contacts are not ideal. As discussed in the next section, this fact can be exploited to observe LL behavior in nanotube devices via the tunneling density of states (TDOS).

3.3 Tunneling Conductance

The fact that a metallic nanotube acts like a nearly perfect 1D conductor with very long mean free path makes it an ideal system to test the LL theory described in Sec. 2. Figure 6 shows the linear-response two-terminal conductance G versus gate voltage V_g for a metallic rope at different temperatures. At low temperatures, the conductance exhibits a series of Coulomb oscillations with a charging energy $E_c = 1.9$ meV. For $k_B T > E_c$, i.e. $T > 20K$, the Coulomb oscillations are nearly completely washed out, and the conductance is independent of gate voltage. A plot of the conductance vs. temperature in this regime is shown in the inset. The conductance drops steeply as the temperature is lowered, extrapolating to $G = 0$ at $T = 0$.

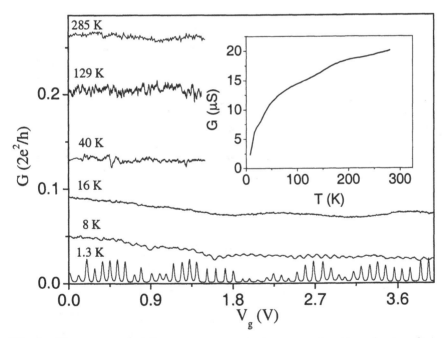

Fig. 6. The two-terminal linear-reponse conductance as a function of gate voltage at a variety of temperatures. The inset shows the average conductance as a function of temperature. Adapted from Ref. [15].

Results for two samples are shown in Fig. 7(a), where the conductance as a function of temperature is plotted on a double-logarithmic scale (solid curves).

Charging effects contribute to the measured characteristics, especially at lower temperatures, $k_B T < 2E_c$. We therefore correct the $G(T)$ data for charging effects by dividing the measured conductance by the theoretically expected temperature dependence of G in the Coulomb blockade model [44]. The dashed lines in Fig. 7(a) show the measured G corrected in this manner as a function of temperature. Looking at the corrected data, we see that they have a finite slope, indicating an approximate power-law dependence upon temperature with exponents $\alpha = 0.33$ and 0.38.

Figure 7(b) shows the measured differential conductance as a function of the applied bias V. The upper left inset to Fig. 7(b) shows $G = dI/dV$ versus V at different temperatures, plotted on a double-logarithmic scale. At low bias, dI/dV is proportional to a (temperature-dependent) constant. At high bias, dI/dV increases with increasing V. The curves at different temperatures fall onto a single curve in the high-bias regime. Since this curve is roughly linear on the double-logarithmic plot, the differential conductance is well described by a power law, $dI/dV \propto V^\alpha$, where $\alpha = 0.36$. At the lowest temperature $T = 1.6$ K, this power-law behavior extends over two decades in the applied voltage V, namely from 1 mV up to 100 mV.

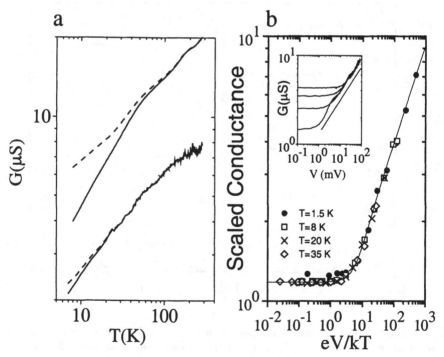

Fig. 7. (a) Conductance plotted against temperature on a double-logarithmic scale for two samples. (b) The scaled differential conductance measured at different temperatures. Adapted from Ref. [15].

Let us next discuss possible origins of this behavior. The data demonstrates that tunneling into the rope has a significant dependence on energy that cannot be described by the Coulomb blockade model. One simple explanation for such behavior is that the transmission coefficients of tunnel barriers are strongly energy-dependent, with substantially increased transparency on the energy scale of the measurement. This would lead e.g. to activated transport over the barrier, $G \propto \exp(-\Delta/k_B T)$. However, the fact that the temperature dependence extrapolates to $G = 0$ at $T = 0$, see inset of Fig. 6, is inconsistent with this functional form. The origin of this behavior appears to originate rather from the TDOS of a LL which vanishes as a power law with energy, see Eq. (24). The assumption that the conductance is limited by tunneling between the metal electrodes and the LL directly leads to the power-law temperature dependence $G(T) \propto T^\alpha$ at small bias, $eV \ll k_B T$. Similarly, for $eV \gg k_B T$, LL theory predicts that $G(V) \propto V^\alpha$. The exponent α follows from the corresponding (bulk-tunneling, see below) exponent in the TDOS. We therefore obtain $\alpha_{\text{bulk}} \approx 0.3$.

The devices used here were made in the following way. Electron beam lithography is first used to define leads, and ropes are deposited on top of the leads. Samples were selected that showed Coulomb blockade behavior at low temperatures with a single well-defined period, indicating the presence of a single dot. The charging energy of these samples indicates a dot with a size substantially larger than the spacing between the leads [22]. Transport thus occurs by electrons tunneling into the middle ("bulk") of the nanotubes. These devices are referred to as "bulk-contacted." We can also use a second method, which was described in Sec. 3.1, where the contacts are applied over the top of the nanotube rope. From measurements of these devices in the Coulomb blockade regime [23], we conclude that the electrons are confined to the length of the rope between the leads. This implies that the leads cut the nanotubes into segments, and transport involves tunneling into the ends of the nanotubes. This type of device is referred to as "end-contacted." For end-contacted devices, similar temperature and voltage dependences of the conductance $G(V, T)$ are observed. The obtained exponent $\alpha_{\text{end}} \approx 0.6$ is significantly larger than $\alpha_{\text{bulk}} \approx 0.3$, the exponent obtained for bulk-contacted devices.

The exponent of these power laws obviously depends on whether the electron tunnels into the end or the bulk of the LL. These exponents are related to the LL parameter g by Eq. (25) and (26), respectively. Using the expected LL parameter $g_{\text{th}} = 0.28$, see Sec. 2.3, the expected exponents are $\alpha_{\text{end,th}} = 0.65$ and $\alpha_{\text{bulk,th}} = 0.24$. The approximate power-law behavior as a function of T or V observed in Fig. 7 follows the theory for tunneling into a LL. The predicted values of the exponents are in good agreement with the experimental values. Remarkably, power-law behavior in T is observed up to 300 K, indicating that nanotubes are LLs even at room temperature.

LL theory makes an additional prediction for this system. Since the temperature and the voltage play an analogous role in the theory, the differential conductance for a single tunnel junction should obey the universal scaling form (27), together with a convolution of the derivative of the Fermi distribution, see

Sec. 2.3. Hence it should be possible to collapse the data onto a single universal scaling curve. To do this, the measured nonlinear conductance $G(V, T) = dI/dV$ at each temperature was divided by T^α and plotted against $eV/k_B T$, as shown in the main body of Fig. 7(b). The data collapses well onto a universal curve. The solid line in Fig. 7(b) is the theoretical plot, see Ref. [15] for details. The theory fits the scaled data quite well.

Recently, Yao et al. [16] have reported on electrical transport measurements on SWNTs with intramolecular junctions. Two nanotubes are connected together by a kink, which acts as a tunnel barrier. In the case of a metal-metal junction, the conductance displays a power-law dependence on temperatures and voltage, consistent with tunneling between the ends of two LLs. The tunneling conductance G across the junction is proportional to the product of the end-tunneling DOS on both sides. Therefore G still varies as a power law of energy, but with an exponent twice as large, namely $\alpha_{end-end} = 2\alpha_{end}$.

4   Discussion and Open Problems

In this review, we have discussed our recent observation of Luttinger liquid behavior in transport experiments on individual metallic carbon nanotubes, along with the detailed theoretical description of this non-Fermi liquid state. The situation in SWNTs seems rather clear by now, since the ballistic nature of transport can be unambiguously established. Nevertheless, several interesting open questions remain. One proposed experiment could probe spin-charge separation by measuring the $I - V$ characteristics of a SWNT in contact to two ferromagnetic reservoirs with continuously varying angle ϕ between the magnetization directions. Another interesting issue concerns the experimental observation of Friedel oscillations in nanotubes, i.e. density oscillations in the conduction electron density around impurities or the end of the tube. These density oscillations should decay with a slow interaction-dependent power law (slower than $1/x$), and, interestingly, the oscillation period can depend on the interaction strength [14]. Furthermore, it is of importance to achieve a better understanding of conduction electron spin resonance (CESR) in SWNTs. Previous experimental attempts have not seen any ESR peak, and one of the proposed reasons for its absence involves electron-electron interactions [45]. However, to the best of our knowledge, there are no theoretical investigations concerning CESR for Luttinger liquids including the gapless charge degrees of freedom. Finally, the phonon backscattering correction to the conductance arising for long SWNTs at high temperatures should involve an anomalous $T^{(1+g)/2}$ scaling [46] that remains to be seen experimentally.

Another line of research currently deals with multi-wall nanotubes which are known to exhibit diffusive transport. Nevertheless, the TDOS apparently shows very similar behaviors as in a SWNT, and superficially it appears that Luttinger liquid concepts also apply to MWNTs. The reason for this is presently unclear, and more theoretical and experimental studies will be needed to clarify the situation.

We acknowledges support by the DFG under the Gerhard-Hess program, by the DOE (Basic Energy Sciences, Materials Sciences Division, the sp2 Materials Initiative), and by DARPA (Moletronics Initiative).

References

1. S. Tomonaga: Prog. Theor. Phys. (Kyoto) **5**, 544 (1950)
2. J.M. Luttinger: J. Math. Phys. (N.Y.) **4**, 1154 (1963)
3. See, e.g., A.A. Abrikosov, L.P. Gorkov, I.E. Dzyaloshinskii: *Methods of Quantum Field Theory in Statistical Physics* (Dover, New York, 1963)
4. F.D.M. Haldane: J. Phys. C **14**, 2585 (1981); Phys. Rev. Lett. **47**, 1840 (1981)
5. See, e.g., A.O. Gogolin, A.A. Nersesyan, A.M. Tsvelik: *Bosonization and Strongly Correlated Systems* (Cambridge University Press, 1998), and J. v. Delft, H. Schoeller: Ann. Phys. (Leipzig) **7**, 225 (1998)
6. M.P.A. Fisher, L.I. Glazman: in *Mesoscopic Electron Transport*, edited by L. Sohn et al. (Kluwer Academic Publishers, 1997)
7. K.V. Pham, M. Gabay, P. Lederer: Phys. Rev. B **61**, 16397 (2000)
8. For a recent review, see J. Voit: Rep. Prog. Phys. **57**, 977 (1995). See also: A. Schwartz, M. Dressel, G. Grüner, V. Vescoli, L. Degiorgi, T. Giamarchi: Phys. Rev. B **58**, 1261 (1998)
9. S. Tarucha, T. Honda, T. Saku: Sol. State Comm. **94**, 413 (1995)
10. O.M. Auslaender, A. Yacoby, R. de Picciotto, K.W. Baldwin, K.W. West: Phys. Rev. Lett. **84**, 1764 (2000)
11. F.P. Milliken, C.P. Umbach, R.A. Webb: Sol. State Comm. **97**, 309 (1996)
12. A.M. Chang, L.N. Pfeiffer, K.W. West: Phys. Rev. Lett. **77**, 2538 (1996); M. Grayson, D.C. Tsui, L.N. Pfeiffer, K.W. West, A. Chang: Phys. Rev. Lett. **80**, 1062 (1998)
13. R. Egger, A.O. Gogolin: Phys. Rev. Lett. **79**, 5082 (1997); C.L. Kane, L. Balents, M.P.A. Fisher: Phys. Rev. Lett. **79**, 5086 (1997)
14. R. Egger, A.O. Gogolin: Eur. Phys. J B **3**, 281 (1998)
15. M. Bockrath, D.H. Cobden, J. Lu, A.G. Rinzler, R.E. Smalley, L. Balents, P.L. McEuen: Nature **397**, 598 (1999)
16. Z. Yao, H.W.J. Postma, L. Balents, C. Dekker: Nature **402**, 273 (1999)
17. C. Biagini, D.L. Maslov, M.Yu. Reizer, L.I. Glazman: preprint cond-mat/0006407
18. R. Mukhopadhyay, C.L. Kane, T.C. Lubensky: preprint cond-mat/0007039
19. S. Iijima: Nature **354**, 56 (1991)
20. Special issue on Carbon Nanotubes in Physics World (June 2000), see especially articles by P. McEuen and by C. Schönenberger and L. Forró
21. C. Dekker: Physics Today (May 1999), p. 22
22. S.J. Tans, M.H. Devoret, H. Dai, A. Thess, R.E. Smalley, L.J. Geerligs, C. Dekker: Nature **386**, 474 (1997); S.J. Tans, M.H. Devoret, R.J.A. Groeneveld, C. Dekker: Nature **394**, 761 (1998)
23. M. Bockrath, D. Cobden, P. McEuen, N. G. Chopra, A. Zettl, A. Thess, R.E. Smalley: Science **275**, 1922 (1997); D.H. Cobden, M. Bockrath, P.L. McEuen, A.G. Rinzler, R. Smalley: Phys. Rev. Lett. **81**, 681 (1998)
24. A. Bachtold, M. Fuhrer, S. Plyasunov, M. Forero, E.H. Anderson, A. Zettl, P.L. McEuen: Phys. Rev. Lett. **84**, 6082 (2000)
25. C.T. White, T.N. Todorov: Nature **393**, 240 (1998)

26. A. Bachtold, C. Strunk, J.P. Salvetat, J.M. Bonard, L. Forró, T. Nussbaumer, C. Schönenberger: Nature **397**, 673 (1999); C. Schönenberger, A. Bachtold, C. Strunk, J.P. Salvetat, L. Forró: Appl. Phys. A **69**, 283 (1999)
27. J. Kim, K. Kang, J.O. Lee, K.H. Yoo, J.R. Kim, J.W. Park, H.M. So, J.J. Kim: preprint cond-mat/0005083
28. R. Egger: Phys. Rev. Lett. **83**, 5547 (1999)
29. S. Frank, P. Poncharal, Z.L. Wang, W.A. de Heer: Science **280**, 1744 (1998)
30. J. van den Brink, G.A. Sawatzky: Europhys. Lett. **50**, 447 (2000)
31. A. Komnik, R. Egger: Phys. Rev. Lett. **80**, 2881 (1998)
32. M.S. Fuhrer, J. Nygard, L. Shih, M. Forero, Y.G. Yoon, M.S.C. Mazzoni, H.J. Choi, J. Ihm, S.G. Louie, A. Zettl, P. McEuen: Science **288**, 494 (2000)
33. R. Egger, H. Grabert: Phys. Rev. Lett. **77**, 538 (1996); **80**, 2255(E) (1998)
34. C.L. Kane, M.P.A. Fisher: Phys. Rev. B **46**, 15233 (1992)
35. R. Egger, H. Grabert, A. Koutouza, H. Saleur, F. Siano: Phys. Rev. Lett. **84**, 3682 (2000)
36. L. Balents, R. Egger: preprint cond-mat/0003038
37. K. Tsukagoshi, B.W. Alphenaar, H. Ago: Nature **401**, 572 (1999)
38. A. Brataas, Yu. Nazarov, G.E.W. Bauer: Phys. Rev. Lett. **84**, 2481 (2000)
39. G.A. Prinz: Physics Today **48(4)**, 58 (1995)
40. Y. Martin, D.W. Abraham, H.K. Wickramasinghe: Appl. Phys. Lett. **52**, 1103 (1988); J.E. Stern *et al.*: Appl. Phys. Lett. **53**, 2717 (1988); C. Schönenberger, S.F. Alvarado: Phys. Rev. Lett. **65**, 3162 (1990)
41. Z. Yao, C.L. Kane, C. Dekker: Phys. Rev. Lett. **84**, 2941 (2000)
42. M.P. Anantram, T.R. Govindan: Phys. Rev. B **58**, 4882 (1998); T. Ando, T. Nakanishi, R. Saito: J. Phys. Soc. Jpn. (Japan) **67**, 2857 (1998)
43. H.T. Soh, A.F. Morpurgo, J. Kong, C.M. Marcus, C.F. Quate, H. Dai: App. Phys. Lett. **75**, 627 (1999)
44. H. Grabert, M.H. Devoret (eds.): *Single Charge Tunneling*, NATO-ASI Series B: Physics, vol. 294 (Plenum Press, New York, 1992)
45. L. Forró, private communcation
46. A. Komnik, R. Egger: in *Electronic Properties of New Materials — Science and Technology of molecular nanostructures*, AIP conference proceedings 486 (Melville, New York, 1999)

Part IV

Phase Coherence

Capacitance, Charge Fluctuations and Dephasing in Coulomb Coupled Conductors

Markus Büttiker

Université de Genève, Département de Physique Théorique,
CH-1211 Genève 4, Switzerland

Abstract. The charge fluctuations of two nearby mesoscopic conductors coupled only via the long range Coulomb force are discussed and used to find the dephasing rate which one conductor exerts on the other. The discussion is based on a formulation of the scattering approach for charge densities and the density response to a fluctuating potential. Coupling to the Poisson equation results in an electrically self-consistent description of charge fluctuations. At equilibrium the low-frequency noise power can be expressed with the help of a charge relaxation resistance (which together with the capacitance determines the RC-time of the structure). In the presence of transport the low frequency charge noise power is determined by a resistance which reflects the presence of shot noise. We use these results to derive expressions for the dephasing rates of Coulomb coupled conductors and to find a self-consistent expression for the measurement time.

1 Introduction

Investigations of time-dependent current fluctuations of mesoscopic systems have been widely used to obtain information which cannot be extracted from conductance measurements alone [1]. In this work we are interested in the fluctuations of the *charge* in a volume element inside the electrical conductor. If the volume element is made very small the fluctuations of interest are thus the fluctuations of the *local* electron density. Such fluctuations can be detected, for instance by measuring the current induced into a nearby gate [2] or a small cavity as shown in Fig. 1. Through the long range Coulomb interaction a charge fluctuation above the average equilibrium value of the charge in the conductor generates additional electric fields which lead to a charge reduction at the surface of the gate. The reduction of charge at the gate surface is accomplished by a flow of carriers out of the contact of the gate [2]. The conductor can be in an equilibrium state in which case the charge fluctuations are associated with Nyquist noise or it can be in a transport state and the charge fluctuations are those that are generated by shot noise.

Charge fluctuations can be detected not only by direct capacitive probing. In an experiment by Buks et al. [3] charge fluctuations are observed through conductance measurements: the charge fluctuations of two conductors in close proximity can give rise to an additional dephasing rate which a carrier in one conductor experiences due to the presence of the other conductor. In the experiment of Buks et al. [3] an Aharonov-Bohm ring with a quantum dot in one of its arms is in close proximity to another conductor which forms a QPC (quantum point contact). The presence of the QPC leads to a reduction of the Aharonov-Bohm interference oscillations. In the experiment of Sprinzak et al. [4] a double quantum dot is brought into the proximity of a conductor in a high magnetic field and the broadening of the Coulomb blockade peak due the charge fluctuations in the edge states of the nearby conductor are measured. Theoretical

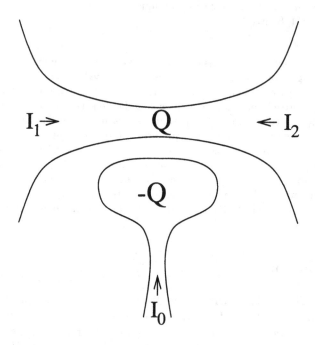

Fig. 1. Cavity with a charge deficit $-Q$ in the proximity of a quantum point contact with an excess charge Q. The dipolar nature of the charge distribution ensures the conservation of currents I_0, I_1 and I_2 flowing into this structure. After [2] .

discussions of dephasing rates in coupled systems are given in Refs. [5–11]. Harris and Stodolsky [5,9] view this as a quantum measurement problem in which the state of one system is measured with the help of another. In this work, as well as in the experimental work of Buks et al. [3] the time it takes to ascertain the state of the measured system, the measurement time, is identified with the dephasing time. Gurvitz investigates the time-evolution of the density matrix of the system that is measured [6]. Aleiner et al. [7] relate the dephasing rate to the orthogonality catastrophe which occurs if an additional carrier is added to the ground state of the system. Levinson derives a dephasing rate in terms of the charge fluctuation spectrum of non-interacting carriers [8]. The approach which we discuss here also relates the dephasing rate to the charge fluctuation spectrum. However, in contrast to the discussions presented in Refs. [5,6,8,9] we emphasize an electrically self-consistent approach which takes into account that charging even in open conductors such as QPC can be energetically expensive [10]. This approach is applicable to a wide range of geometries and in Refs. [10,11] has been used to present a self-consistent treatment of charge fluctuations in edge states. Our discussion can be compared with Ref. [12] which treats fluctuations as a free electron problem.

It is interesting to notice that in discussions of charge fluctuations in systems that are composed entirely of components in which charge quantization is important, theoretical work [13–17] carefully discusses the various capacitance coefficients which determine the charging energies of the system and the coupling. On the other hand, when subsystems like QPC's are discussed, electric carriers are treated as if they were non-interacting entities [14–17]. Clearly, a QPC has also a capacitance. A QPC can be characterized by a capacitance which describes its self-polarization [18–20]. This self-polarization corresponds to charge accumulation on one side of the quantum point contact and charge depletion on the other side of the quantum point contact [18,19]. The self-polarization does not change the overal charge of the QPC and is thus here not of primary importance. But a quantum point contact can be charged vis-a-vis the gates [21] or vis-a-vis any other conductor. This leads to a net charge on the quantum point contact and is the dominant process by which the quantum point contact interacts with an other conductor [21].

Therefore, like in the systems which exhibit Coulomb blockade, we can ask: "What is the dependence of a charge fluctuation spectrum of a QPC on its capacitance?", or "How does the additional dephasing rate generated by the proximity of a QPC depend on the capacitance of the quantum point contact?" If the relevant capacitance were a mere geometrical quantity these questions might just determine some prefactors left open in previous work. However, the charge screening of a QPC is non-trivial since the density of states of a quantum point contact [2] (at least in the semi-classical limit) diverges for a gate voltage at which a new channel is opened (quantum tunneling limits the density of states).

The aim of this work is to present a simple self-consistent discussion of charge fluctuation spectra of Coulomb coupled conductors and to use these spectra to find dephasing rates and self-consistent expression for the measurement time. We assume that the ground state of the system has been determined and investigate small deviations away from the ground state. The approach which we present combines the scattering approach with the Poisson equation and treats interactions in the random phase approximation [10]. This approach has been used with some success to treat the dynamic conductance [22,23] of mesoscopic systems, their non-linear I-V-characteristics [24–26], and higher harmonics generation [25]. Indeed there is a close connection between the charge fluctuation spectra which we obtain and the dynamic conductances of a mesoscopic system [22,21]. We show that the charge noise power of the equilibrium charge fluctuations can at low frequencies be characterized by a charge relaxation resistance R_q. Together with the capacitance this resistance determines the RC-time of the mesoscopic structure [22]. Indeed, independent measurements of the capacitance and the resistance R_q, when compared with the results from a measurement of the dephasing rate would provide an important overall test of the consistency of the theory. In the presence of transport, we deal in the low temperature limit with shot noise. In this case, the low frequency charge noise power is proportional to the applied voltage and proportional to a resistance R_v.

Some works advocate a perturbative treatment of coupled systems and present results which are proportional to the square of the coupling constant. In contrast, the self-consistent approach discussed here leads to a more intersting dependence on the bare coupling constant [10] (in the physically most relevant case we actually find a dephasing rate which is proportional to one over the square of the coupling strength).

In this work, we treat only the symmetric QPC Coulomb coupled to another system. This limitation is motivated by the fact that the QPC is the most widely investigated example. This should permit a most direct comparison of the self-consistent approach advocated here with the results in the literature. For a review which addresses a wider range of geometries we refer the reader to [27]. We also restrict ourselves to the case where linear screening is applicable.

An evaluation of the dephasing rates requires a discussion of the potential fluctuations (or equivalently, the charge fluctuations). We will need only the zero-frequency limit of the potential fluctuation spectrum, and it is thus sufficient to find the zero-frequency, white noise limit of the charge fluctuations.

2   The Effective Interaction

To be definite, consider the conductors of Fig. 1. We first investigate the relation between voltage and charges of these two conductors for small deviations of the applied voltages away from their equilibrium value. We assume that each electric field line emanating from the cavity ends up either again on the cavity or on the QPC. There exists a Gauss volume which encloses both conductors which can be chosen large enough so that the electrical flux through its surface vanishes [23]. Consequently, the charge on the two conductors is conserved. Any accumulation of charge at one location within our Gauss volume is compensated by a charge depletion at another location within the Gauss volume. The variation of the charge brought about by a small change of the applied voltage is thus of a dipolar nature.

To keep the discussion simple, we consider here the case that each conductor is described by a single potential only. The charge and potential on the cavity are denoted by dQ_0 and dU_0 and on the QPC by dQ_1 and dU_1. A more accurate description can be obtained by subdividing the conductor into a number of volume elements [21] or in fact by using a continuum description [23]. The essential elements of our discussion do, however, already become apparent in the simple case that each conductor is described only by one potential and we will treat only this limiting case in this work. The Coulomb interaction is described with the help of a single *geometrical* capacitance C. The charges and potentials of the two conductors are then related by

$$dQ_0 = C(dU_0 - dU_1), \tag{1}$$

$$dQ_1 = C(dU_1 - dU_0). \tag{2}$$

The two equations can be thought off as a discretized version of the Poisson equation. Note that according to Eqs. (1,2) we have $dQ_0 + dQ_1 = 0$.

We now complement these two equations by writing the charges dQ_i as a sum of an external (or bare) charge calculated for fixed internal potentials U_0 and U_1 and an induced charge generated by the response of the potential due to the injected charges. The additional charge injected into the cavity (which we denote by edN_0) due to an increase of the reservoir voltage by edV_0 is $edN_0 = eD_0edV_0$. Here D_0 is the total density of states at the Fermi energy of the cavity of the region in which dU_0 deviates from its equilibrium value. The induced charge is $-e^2 D_0 dU_0$. It is negative since the Coulomb interaction counteracts charging. It is also determined by the total density of states since the integrated Lindhard function is given by the density of states [28]. Thus the charge on the cavity is

$$dQ_0 = (edN_0 - e^2 D_0 dU_0). \tag{3}$$

To find the charge on the QPC we must take into account that there are two reservoir potentials which we denote by dV_1 and dV_2. An increase in the potential of contact of reservoir 1 (at constant internal potential U_1) does not fill all the available states but only a portion. The density of states [21] of carriers incident from reservoir 1 is denoted by D_{11} and is called the *injectance* of contact 1. Thus the charge injected at constant internal potential by an increase of the voltage V_1 is $eD_{11}edV_1$. Similarly an increase of the potential at reservoir 2 leads to an additional charge $eD_{12}edV_2$. Here D_{12} is the injectance of reservoir 2. Together these two (partial) density of states are equal to the total density of states of the quantum point contact $D_{11} + D_{12} = D_1$. Thus the total injected charge on conductor 1 is $edN_1 = eD_{11}edV_1 + eD_{12}edV_2$. Screening is against the total density of states and thus a variation of the internal potential dU_1 generates an induced charge given by $e^2 D_1 dU_1$. The charge dQ_1 is the sum of these three contributions,

$$dQ_1 = (edN_1 - e^2 D_1 dU_1). \tag{4}$$

We arrive thus at the following self-consistent equations relating charges and potentials,

$$dQ_0 = C(dU_0 - dU_1) = (edN_0 - e^2 D_0 dU_0), \tag{5}$$

$$dQ_1 = C(dU_1 - dU_0) = (edN_1 - e^2 D_1 dU_1). \tag{6}$$

We can use these equations to express the internal potentials U_i in terms of the injected charges edN_i. We find $dU_i = e \sum_j G_{ij} dN_j$ with an effective interaction G_{ij} given by

$$\mathbf{G} = \frac{C_\mu}{e^2 D_0 e^2 D_1 C} \begin{pmatrix} C + e^2 D_1 & C \\ C & C + e^2 D_0 \end{pmatrix}. \tag{7}$$

Here C_μ is the electrochemical capacitance

$$C_\mu^{-1} = C^{-1} + (e^2 D_0)^{-1} + (e^2 D_1)^{-1} \tag{8}$$

which is the series capacitance of the geometrical contribution C and the density of states of the two conductors [22]. Note that in contrast to perturbation treatments, the effective coupling element G_{12} is not proportional to e^2/C but in general is a complicated function of this energy. We will use the effective interaction in Section 7 to find the measurement time. First, however, we will now use the effective interaction to express the true charge fluctuations in terms of the bare fluctuations.

We are interested not in the average quantities discussed above but in their dynamic fluctuations. To this extend we now re-write Eqs. (5) and (6) for the fluctuating quantities. In a second quantization approach the fluctuating quantities are described with the help of operators, $\hat{Q}_i$ for the true charges, and the potentials $\hat{U}_i$ on the two conductors, $i = 0, 1$. As for the average charge, the fluctuating charge can also be written in terms of bare charge fluctuations $e\hat{N}_i$ (calculated by neglecting the Coulomb interaction) counteracted by a screening charge $eD_i e\hat{U}_i$. Below, we will give explicit expressions for all these operators. Instead of Eqs. (5) and (6) we now have,

$$\hat{Q}_0 = C(\hat{U}_0 - \hat{U}_1) = e\hat{N}_0 - e^2 D_0 \hat{U}_0, \tag{9}$$

$$\hat{Q}_1 = C(\hat{U}_1 - \hat{U}_0) = e\hat{N}_1 - e^2 D_1 \hat{U}_1. \tag{10}$$

Clearly, if we consider simply the average of these equations, they must reduce to Eq. (5) and (6). The fluctuations are determined by the off-diagonal elements of the charge and potential operators. Below we will specify these expressions in detail. Solving these equations for the potential operators, we find $\hat{U}_i = e \sum_j G_{ij} \hat{N}_j$ with the effective interaction G_{ij} given by Eq. (7).

Let us now introduce the noise power spectra of the bare charges, $S_{N_i N_i}(\omega)$ for each of the conductors. The spectrum of the bare charge fluctuations is defined as

$$S_{N_i N_i}(\omega) 2\pi\delta(\omega + \omega') = \langle \hat{N}_j(\omega)\hat{N}_j(\omega') + \hat{N}_j(\omega')\hat{N}_j(\omega) \rangle \tag{11}$$

with $\hat{N}_j(\omega) = \hat{N}_j(\omega) - \langle \hat{N}_j(\omega) \rangle$, where $\hat{N}_j(\omega)$ is the Fourier transform of the charge operator of conductor j.

The bare charge fluctuation spectra on different conductors are uncorrelated, $S_{N_i N_j}(\omega) = 0$ for $i \neq j$. With the help of the effective interaction matrix, we can now relate the potential fluctuation spectra to the fluctuation spectra of the bare charges. In the zero-frequency limit we find,

$$S_{U_i U_j}(0) = e^2 \sum_k G_{ik} G_{jk} S_{N_k N_k}(0). \tag{12}$$

Even though the bare charge fluctuations are uncorrelated, the potential fluctuations and the true charge fluctuations on the two conductors are correlated.

3   Charge Relaxation Resistances

It is useful to characterize the noise power of the charge fluctuations with the help of resistances. Consider first the case where both conductors are at equilibrium. Charge fluctuations on the conductors arise due to the random thermal injection of carriers. The bare charge fluctuation spectrum, normalized by the density of states D_i of conductor i has the dimension of a resistance. We introduce the charge relaxation resistance $R_q^{(j)}$ of conductor j,

$$2kTR_q^{(j)} \equiv e^2 S_{N_j N_j}(0)/(e^2 D_j)^2. \tag{13}$$

The charge relaxation resistance has a physical significance in a number of problems. In simple cases, R_q together with an appropriate capacitance determines the RC-time of the mesoscopic structure [22]. The charge relaxation resistance can thus alternatively be determined by investigating the poles of the conductance matrix [23,22]. The dynamic conductance matrix $G_{\alpha\beta}(\omega) \equiv dI_\alpha(\omega)/dV_\beta(\omega)$ of our mesoscopic structure (QPC and cavity) which relates the currents $dI_\alpha(\omega)$ at a frequency ω at contact α to the voltages $dV_\beta(\omega)$ applied at contact β has at low frequencies a pole determined by $\omega_{RC} = -iC_\mu(R_q^{(1)} + R_q^{(2)})$. Alternatively we could carry out a low frequency expansion of the element $G_{00}(\omega)$ (the element of the conductance matrix which gives the current at the contact of the cavity in response to an oscillating voltage applied to the cavity) to find [22,23] that $G_{00}(\omega) = -iC_\mu\omega + C_\mu^2 R_q\omega^2 + ...$ Thus R_q plays a role in many problems. The charge relaxation resistance differs from the dc-resistance. For instance a ballistic one-channel quantum wire connecting two reservoirs and capacitively coupled to a gate has for spinless carriers a dc-resistance of $R = h/e^2$ and a charge relaxation resistance [31] of $R_q = h/4e^2$. The dc-resistance corresponds to the series addition of resistances along the conductance path, whereas an excess charge on the conductor relaxes via all possible conductance channels to the reservoirs and thus corresponds to the addition of resistances in parallel. This is nicely illustrated for a chaotic cavity [32] connected via contacts with M_1 and M_2 perfectly transmitting channels to reservoirs and capacitively coupled to a gate. Its ensemble averaged dc-resistance is $R = (h/e^2)(M_1^{-1} + M_2^{-1})$, whereas its charge relaxation resistance is $R_q = (h/e^2)(M_1 + M_2)^{-1}$. Thus the dc-resistance is governed by the smaller of the two contacts, whereas the charge relaxation resistance is determined by the larger contact.

If the conductor is driven out of equilibrium with the help of an applied voltage $|V| \equiv |V_1 - V_2|$, the thermal noise described by Eq. (13) can be overpowered by shot noise. For $e|V| \gg kT$ the charge fluctuation spectrum becomes proportional to the applied voltage and defines a resistance $R_v^{(j)}$ via the relation,

$$2e|V|R_v^{(j)} \equiv e^2 S_{N_j N_j}(0)/(e^2 D_j)^2. \tag{14}$$

The resistance R_v is thus a measure of the noise power of the charge fluctuations associated with shot noise.

Eq. (13) and Eq. (14) describe the behavior of $e^2 S_{N_j N_j}(0)/(e^2 D_j)^2$ in the limits $kT \gg e|V|$ and $kT \ll e|V|$. For fixed temperature as a function of voltage

$e^2 S_{N_j N_j}(0)/(e^2 D_j)^2$ exhibits a smooth crossover from the equilibrium result Eq. (13) to Eq. (14) valid in the presence of shot noise. For the structure shown in Fig. 1 it is only the QPC (conductor 1) which can be brought out of equilibrium. The cavity, connected to a single lead always exhibits only thermal fluctuations and its charge relaxation resistance is characterized by R_q^0 even if the QPC is subject to shot noise.

4   Bare Charge Fluctuations and the Scattering Matrix

Let us now determine the charge operator for the bare charges (non-interacting carriers). The operator for the total charge on a mesoscopic conductor can be found from the current operator and by integrating the continuity equation over the total volume of the conductor. This gives a relation between the charge in the volume and the particle currents entering the volume. We obtain for the density operator [2]

$$\hat{N}(\omega) = \hbar \sum_{\beta\gamma} \sum_{mn} \int dE \, \hat{a}^{\dagger}_{\beta m}(E) \mathcal{D}_{\beta\gamma mn}(E, E + \hbar\omega) \hat{a}_{\gamma n}(E + \hbar\omega), \qquad (15)$$

where $\hat{a}^{\dagger}_{\beta m}(E)$ (and $\hat{a}_{\beta m}(E)$) creates (annihilates) an incoming particle with energy E in lead β and channel m. The element $\mathcal{D}_{\beta\gamma mn}(E, E + \hbar\omega)$ is the non-diagonal density of states element generated by carriers incident simultaneously in contact β in quantum channel m and by carriers incident in contact γ in channel n. In particular, in the zero-frequency limit, we find in matrix notation [2,10],

$$\mathcal{D}_{\beta\gamma}(E) = \frac{1}{2\pi i} \sum_{\alpha} \mathbf{s}^{\dagger}_{\alpha\beta}(E) \frac{d\mathbf{s}_{\alpha\gamma}(E)}{dE}. \qquad (16)$$

Expressions of this type are known from the discussion of quantum mechanical time delay [33]. The sum of the diagonal elements of this matrix is the density of states of the conductor

$$D(E) = \sum_{\beta} Tr[\mathcal{D}_{\beta\beta}(E)] = \frac{1}{2\pi i} \sum_{\alpha,\beta} Tr[\mathbf{s}^{\dagger}_{\alpha\beta}(E) \frac{d\mathbf{s}_{\alpha\beta}(E)}{dE}], \qquad (17)$$

where the trace is over the quantum channels. $\mathcal{D}_{\beta\beta} \equiv \mathcal{D}_{\beta}$ is the injectance of contact β. (We used this density of states in the discussion leading to Eq. (3)).

The charge fluctuations are determined by the off-diagonal elements of Eq. (15). Proceeding as for the case of current fluctuations [34,35,1] we find for the fluctuation spectrum of the total charge

$$S_{NN}(0) = 2h \sum_{\gamma\delta} \int dE \, Tr[\mathcal{D}^{\dagger}_{\gamma\delta}\mathcal{D}_{\delta\gamma}] f_{\gamma}(E)(1 - f_{\delta}(E)). \qquad (18)$$

The spectrum of the bare charge fluctuations has to be determined for each conductor separately using its scattering matrix. We now go on to find specific expression for this spectrum for the QPC.

5 Charge Relaxation Resistance of a Quantum Point Contact

To illustrate the preceding discussion, we now consider specifically the charge relaxation resistance and subsequently the resistance R_v of a QPC. For simplicity, we consider a symmetric QPC (the asymmetric case [4] is treated in [10,12]): For a symmetric scattering potential the scattering matrix (in a basis in which the transmission and reflection matrices are diagonal) is for the n-th channel of the form

$$s_n(E) = \begin{pmatrix} -i\sqrt{R_n}\exp(i\phi_n) & \sqrt{T_n}\exp(i\phi_n) \\ \sqrt{T_n}\exp(i\phi_n) & -i\sqrt{R_n}\exp(i\phi_n) \end{pmatrix}, \tag{19}$$

where T_n and $R_n = 1 - T_n$ are the transmission and reflection probabilities and ϕ_n is the phase accumulated by a carrier in the n-th eigen channel. We find for the elements of the density of states matrix, Eq. (16),

$$\mathcal{D}_{11} = \mathcal{D}_{22} = \frac{1}{2\pi}\frac{d\phi_n}{dE}, \quad \mathcal{D}_{12} = \mathcal{D}_{21} = \frac{1}{4\pi}\frac{1}{\sqrt{R_n T_n}}\frac{dT_n}{dE}. \tag{20}$$

With these density of states matrix elements, we can determine the particle fluctuation spectrum, Eq. (18) in the white-noise limit, and R_q with the help of Eq. (13)

$$R_q = \frac{h}{4e^2}\frac{\sum_n\left[(\frac{d\phi_n}{dE})^2 + \frac{1}{4T_nR_n}(\frac{dT_n}{dE})^2\right]}{[\sum_n\frac{d\phi_n}{dE}]^2}. \tag{21}$$

Eq. (21) is still a formal result, applicable to any symmetric (two-terminal) conductor. To proceed we have to adopt a specific model for a QPC. If only a few channels are open the average potential has in the center of the conduction channel the form of a saddle [36]:

$$V_{eq}(x,y) = V_0 + \frac{1}{2}m\omega_y^2 y^2 - \frac{1}{2}m\omega_x^2 x^2, \tag{22}$$

where V_0 is the potential at the saddle and the curvatures of the potential are parametrized by ω_x and ω_y. The resulting transmission probabilities have the form of Fermi functions $T(E) = 1/(e^{\beta(E-\mu)} + 1)$ (with a negative temperature $\beta = -2\pi/\hbar\omega_x$ and $\mu = \hbar\omega_y(n + 1/2) + V_0$). As a function of energy (gate voltage) the conductance rises step-like. The energy derivative of the transmission probability $dT_n/dE = (2\pi/\hbar\omega_x)T_n(1 - T_n)$ is itself proportional to the transmission probability times the reflection probability. We note that such a relation holds not only for the saddle point model of a QPC but also for instance for the adiabatic model [37]. As a consequence $(1/4T_nR_n)(dT_n/dE)^2 = (\pi/\hbar\omega_x)^2 T_n R_n$ is proportional to $T_n R_n$. Thus the charge relaxation resistance of a saddle QPC is

$$R_q = \frac{h}{4e^2}\frac{\sum_n\left[(\frac{d\phi_n}{dE})^2 + (\frac{\pi}{\hbar\omega_x})^2 T_n R_n\right]}{[\sum_n\frac{d\phi_n}{dE}]^2}. \tag{23}$$

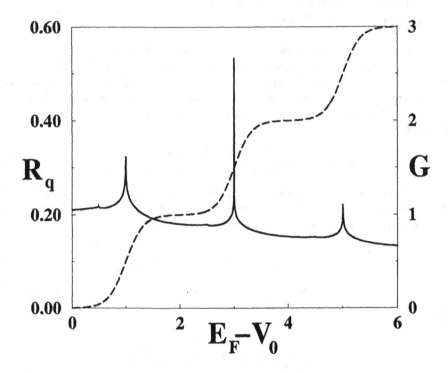

Fig. 2. Charge relaxation resistance R_q of a saddle QPC in units of $h/4e^2$ for $\omega_y/\omega_x = 2$ and a screening length of $m\omega_x\lambda^2/\hbar = 2E_\lambda/\hbar\omega_x = 25$ as a function of $E_F - V_0$ in units of $\hbar\omega_x$ (full line). The broken line shows the conductance of the QPC. After [10] .

To find the density of states of the n-th eigen channel, we use the relation between density and phase (action) and $D_n = \sum_i \mathcal{D}_{n,ii} = (1/\pi)d\phi_n/dE$. We evaluate the phase semi-classically. The spatial region of interest for which we have to find the density of states is the region over which the electron density in the contact is not screened completely. We denote this length by λ and the associated energy by $E_\lambda = (1/2)m\omega_x^2\lambda^2$. The density of states is then found from $D_n = 1/h \int_{-\lambda}^{\lambda} \frac{dp_n}{dE} dx$ where p_n is the classically allowed momentum. A calculation gives a density of states [2] $D_n(E) = (4/(h\omega_x)) \operatorname{asinh}[E_\lambda/(E - E_n)]^{1/2}$, for energies E exceeding the channel threshold E_n and gives a density of states $D_n(E) = (4/(h\omega_x)) \operatorname{acosh}[E_\lambda/(E_n - E)]^{1/2}$ for energies in the interval $E_n - E_\lambda \leq E < E_n$ below the channel threshold. Electrons with energies less than $E_n - E_\lambda$ are reflected before reaching the region of interest, and thus do not contribute to the density of states. The resulting density of states has a logarithmic singularity at the threshold $E_n = \hbar\omega_y(n + \frac{1}{2}) + V_0$ of the n-th quantum channel. (A fully quantum mechanical calculation gives a density of states which exhibits also a peak at the threshold but which is not singular).

We now have all the elements to calculate the charge relaxation resistance R_q and the resistance R_v. The charge relaxation resistance for a saddle QPC is shown in Fig. 2 for a set of parameters given in the figure caption. The charge relaxation resistance exhibits a sharp spike at each opening of a quantum channel. Physically this implies that the relaxation of charge, determined by the RC-time is very rapid at the opening of a quantum channel.

Similarly, we can find the resistance R_v for a QPC subject to a voltage $e|V| >> kT$. Using the density matrix elements for a symmetric QPC given by Eqs. (20), we find [2]

$$R_v = \frac{h}{e^2} \frac{\sum_n \frac{1}{4R_n T_n} \left(\frac{dT_n}{dE}\right)^2}{[\sum_n (d\phi_n/dE)]^2} = \frac{h}{e^2} \left(\frac{\pi}{\hbar\omega_x}\right)^2 \frac{\sum_n T_n R_n}{[\sum_n (d\phi_n/dE)]^2}. \tag{24}$$

The resistance R_v is shown in Fig. 3.

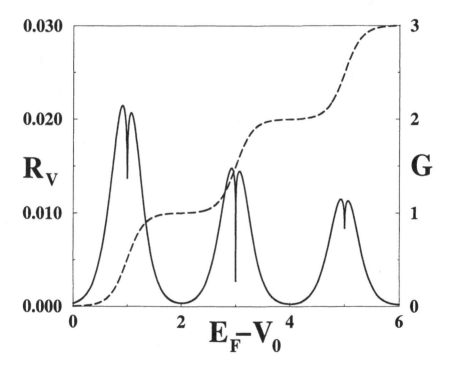

Fig. 3. R_v (solid line) for a saddle QPC in units of h/e^2 and G (dashed line) in units of e^2/h as a function of $E_F - V_0$ in units of $\hbar\omega_x$ with $\omega_y/\omega_x = 2$ and a screening length $m\omega_x\lambda^2/\hbar = 25$. R_v and G are for spinless electrons.

6   Dephasing Rates and Potential Fluctuations

Let us now use the results discussed above to find the dephasing rates in the Coulomb coupled conductors. Consider a scattering state $\Psi_i(\mathbf{r}, E)$ at energy E in conductor i which solves the Schrödinger equation for a fixed potential $U_{eq,i}(\mathbf{r})$. Fluctuations of the potential $U_i(\mathbf{r}, t)$ away from the static (average) equilibrium potential will scatter the carrier out of the eigenstate $\Psi_i(\mathbf{r}, E)$. Here we regard the fluctuating potential in the interior of the conductor as spatially uniform. Thus the fluctuating potential $dU_i(t) = U_i(t) - U_{eq,i}$ in the region of interest is a function of time only. The effect of the fluctuating potential can then be described with the help of a time-dependent phase $\phi(t)$ which multiplies the scattering state. Thus we consider a solution of the type $\Psi_i(\mathbf{r}, E)\exp(-i\phi(t))$ of the time dependent Schrödinger equation. The equation of motion for the phase is simply, $\hbar d\phi/dt = edU_i(t)$. Let us characterize the potential fluctuations in conductor i by its noise spectrum $S_{U_iU_i}(\omega)$. Using the noise spectrum $S_{U_iU_i}(\omega)$ of the voltage fluctuations we find that at long times the phase ϕ of the scattering state diffuses with a rate

$$\Gamma_\phi^{(i)} = \langle(\phi(t) - \phi(0))^2\rangle/2t = (e^2/2\hbar^2)S_{U_iU_i}(0) \tag{25}$$

determined by the zero-frequency limit of the noise power spectrum of the potential fluctuations.

Eq. (12) shows that the potential fluctuations and thus the dephasing rate has two sources: A carrier in conductor i suffers dephasing due to charge fluctuations in conductor $j = i$ itself, and due to charge fluctuations of the additional nearby conductor $j \neq i$. Accordingly, we can also write the dephasing rate in conductor i as a sum of two contributions, $\Gamma_\phi^{(i)} = \sum_j \Gamma_\phi^{(ij)}$ with

$$\Gamma_\phi^{(ij)} = (e^4/2\hbar^2)G_{ij}^2 S_{N_jN_j}(0). \tag{26}$$

At equilibrium, we can express the charge noise power with the help of the equilibrium charge relaxation resistances R_q^j given by Eq. (13). Of particular interest is the dephasing rate in the cavity (conductor $i = 0$) due to the presence of the QPC (conductor $i = 1$). The self-consistent theory gives for this dephasing rates at equilibrium [10]

$$\Gamma_\phi^{(01)} = (e^2/\hbar^2)(e^2D_1G_{01})^2 R_q^{(1)}kT. \tag{27}$$

In the non-equilibrium case, if conductor 1 is subject to a current generated by a voltage $|V|$ (with $kT << e|V|$) we find a dephasing rate [10]

$$\Gamma_\phi^{(01)} = (e^2/\hbar^2)(e^2D_1G_{01})^2 R_v^{(1)}e|V|. \tag{28}$$

These dephasing rates are proportional to $(e^2D_1G_{01})^2$. Thus $e^2D_1G_{01}$ plays the role of an effective coupling constant in our problem. In the limit of a small Coulomb energy ($C \gg e^2D_0$ and $C \gg e^2D_1$) we find $e^2D_1G_{01} = D_0/(D_0 + D_1)$ *independent* of the geometrical capacitance C. In the limit of a large Coulomb energy ($C \ll e^2 D_0$ and $C \ll e^2 D_1$) the effective coupling constant becomes proportional to the capacitance $e^2 D_1 G_{01} = C/(e^2 D_0)$. The second limit, typically, is the physically relevant limit.

According to Eq. (27) and Eq. (28), in an experiment in which the QPC is opened with the help of a gate, the dephasing rate follows, at equilibrium just R_q and in the non-equilibrium case follows R_v. Without screening R_v would exhibit a bell shaped behavior as a function of energy, i. e. it would be proportional to $T_n(1-T_n)$ in the energy range in which the n-th transmission channel is partially open. Screening, which in R_v is inversely proportional to the density of states squared, generates the dip at the threshold of the new quantum channel at the energy which corresponds to $T_n = 1/2$ (see Fig. 3). It is interesting to note that the experiment [3] does indeed show a double hump behavior of the dephasing rate.

7 Self-Consistent Measurement Time

Suppose that a measurement of the current is used to determine the charging state of the cavity. Consider the two charging states with Q_0 and Q_0+e electrons on the cavity. The two charge states on the cavity give, via long range Coulomb interaction, rise to a conductance G (charge Q_0) and $G + \Delta G$ (charge $Q_0 + e$). The measurement time τ_m is the (minimal) time needed to determine the conductance through a measurement that flows through the QPC. A measurement needs to overcome the fluctuations of the current (shot noise and thermal noise). The measurements needs to be long enough [7,3] so that the integrated current $\Delta G|V|\tau_m$ due to the variation of the conductance, exceeds the integrated current fluctuation $\sqrt{S_{II}(0)\tau_m}$. Here, $S_{II}(0)$ is the low frequency spectral current density, which in the zero temperature limit is due to shot nose alone and given by [34,35,1] $S_{II}(0) = 2e(e^2/h)|V| \sum_n T_n R_n$. This gives a measurement time

$$\tau_m = \frac{S_{II}}{(\Delta G)^2 |V|^2}. \tag{29}$$

ΔG is another quantity that can be measured independently and such a measurement has in fact been carried out by Buks et al. [3]. It is thus useful to make a theoretical prediction for this quantity. Therefore our purpose here is to evaluate ΔG and compare the expression for the measurement time with the phase breaking rate obtained above.

The variation of the conductance of the QPC is determined by the sensitivity of the conductance due to the variation of the potential dU_1 in the QPC

$$\Delta G = (e^2/h)(dT/dU_1)dU_1. \tag{30}$$

Here T is the total transmission probability, $T \equiv \sum_n T_n$. In Eq. (30) U_1 is the potential in the QPC and dU_1 is the change in potential for the case that an additional electron eners the cavity. In WKB-approximation we are allowed to replace the derivative with respect to the potential with a derivative with respect to energy, $dT/dU_1 = -edT/dE$. For the saddle point QPC we have $dT/dE = (2\pi/\hbar\omega_x)\sum_n T_n R_n$. Form Eq. (5) we find with $edN_0 = e$, a potential variation $dU_1 = G_{10}edN_0 = eG_{10}$ where G_{10} is an off-diagonal element of the effective interaction matrix, Eq. (7). Thus the addition of an electron onto the cavity changes the conductance by [38]

$$\Delta G = -(e^2/h)eG_{10}dT/dE. \tag{31}$$

Since $G_{10} = C_\mu/(e^2 D_0 e^2 D_1)$ we find in the limit $e^2/C >> 1/D_1$, $e^2/C >> 1/D_2$,

$$\Delta G = -(e^2/h)(C/e^2 D_0)(e(dT/dE)/e^2 D_1). \tag{32}$$

As discussed above, as a function of gate voltage the density of states D_1 of the QPC exhibits a strong variation. In particular, at zero temperature, the density of states diverges at the threshold of a new quantum channel. Consequently, ΔG is also a strong function of gate voltage. ΔG vanishes on the conductance plateaus since (dT/dE) vanishes on a plateau. Eq. (31) predicts that ΔG vanishes also at the channel opening threshold ($T_n = 1/2$) since the semiclassical density of states diverges. Thus Eq. (31) predicts that ΔG (like R_v) is maximal away from the channel opening threshold. Such a behavior is not seen in the experiment of Buks et al. [3], ΔG seems to be rather independent of the gate voltage. This can be due to the simple model of the QPC used here or due to the fact that the experiment is not in the zero-temperature limit.

We now return to the measurement time, Eq. (29). Using Eq. (31) and $dT/dE = (2\pi/\hbar\omega_x)\sum_n T_n R_n$, and the expression for R_v as given by Eq. (24) we find,

$$\tau_m = \frac{2e^2}{\pi^2(e^2/h)^2[e^2 G_{10}D_1]^2 R_v e|V|}. \tag{33}$$

Comparison with our result for Γ_ϕ (see Eq. (28)) shows that $(1/2)\Gamma_\phi\tau_m = 1$ which agress with Korotkov [17]. As mentioned in the introduction, the identification of the measurement time τ_m with $1/\Gamma_\phi$ is taken for granted by a several authors. It is now, however, clear that in general [13,17] $(1/2)\Gamma_\phi\tau_m > 1$ (for instance for a detector that is not symmetric). Even for the symmetric detector (symmetric QPC) considered here, it is clear that a non-zero temperature has a different effect on the dephasing time and on the measurement time. The dephasing time $\tau_\phi = 1/\Gamma_\phi$ is inversely proportional to the charge fluctuation spectrum whereas the measurement time is proportional to the current fluctuation spectrum. At elevated temperatures, the measurement must overcome the combined thermal and shot noise and the measurement time will thus increase with increasing temperature. On the other hand the additional Nyquist noise leads to a shorter dephasing time.

8 Discussion

We have presented an electrically self-consistent discussion of dephasing rates and measurement times for Coulomb coupled conductors. The approach emphasizes that also in open conductors, like QPC's, charge fluctuations are associated with a Coulomb energy. In such a self-consistent treatment, the dephasing rates are typically not simply proportional to a coupling constant. In this work, we have attributed only a single potential to each conductor, but the theory [10] is not in fact limited to such a simplification and permits the treatment of an arbitrary potential landscape [10,39]. The theory also permits a discussion of a wide variety of geometries [39,27].

We have treated the charge fluctuations within a linear screening approach. Large changes in the potential of the QPC would require a discussion of non-linear screening. In either case, a theory is necessary which treats the true charge distribution and its fluctuations. A carrier on the cavity is entangled not only with a single electron on the QPC but with all electrons which are involved in the screening process on the cavity and on the QPC. Instead of a few electron problem our approach emphasizes the true many body nature of charge fluctuations of Coulomb coupled conductors. We have restricted our considerations to the case where we deal with open conductors for which charge quantization plays a minor role. The considerations given above apply, however, also to the case where we have a QPC interacting with a system in which charge is quantized. Even in this case carriers on the QPC will be screened to a certain extent and the charge relaxation resistance R_q and the resistance R_v should again be part of a self-consistent answer.

Acknowledgement

This work was supported by the Swiss National Science Foundation and the TMR network.

References

1. Ya. M. Blanter and M. Büttiker, Physics Reports, **336**, 1-166 (2000).
2. M. H. Pedersen, S. A. van Langen and M. Büttiker, Phys. Rev. B **57**, 1838 (1998).
3. E. Buks, R. Schuster, M. Heiblum, D. Mahalu and V. Umansky, Nature **391**, 871 (1998).
4. D. Sprinzak, E. Buks, M. Heiblum and H. Shtrikman, Phys. Rev. Lett. **84**, 5820 (2000).
5. R. A. Harris and L. Stodolsky, J. Chem. Phys. **74**, 2145 (1981); Phys. Lett. B**116**, 464 (1982).
6. S. A. Gurvitz, Phys. Rev. B **56**, 15215 (1997).
7. I. L. Aleiner, N. S. Wingreen, and Y. Meir, Phys. Rev. Lett. **79**, 3740 (1997).
8. Y. B. Levinson, Europhys. Lett. **39**, 299 (1997).
9. L. Stodolsky, Phys. Lett. B **459**, 193 (1999).
10. M. Büttiker and A. M. Martin, Phys. Rev. B**61**, 2737 (2000).

11. M. Büttiker, in *Statistical and Dynamical Aspects of Mesoscopic Systems*, edited by D. Reguera, G. Platero, L. L. Bonilla and J. M. Rubi, (Springer, Berlin, 2000). Lecture Notes in Physics, Vol. 547, p. 81.

12. Y. B. Levinson, Phys. Rev. B **61**, 4748 (2000).

13. A. Shnirman and G. Schön, Phys. Rev. B**57**, 15400 (1998).

14. Y. Makhlin, G. Schön, A. Shnirman, (unpublished). cond-mat/0001423

15. A. N. Korotkov, D. V. Averin, (unpublished). cond-mat/0002203

16. D.V. Averin. In *Exploring the Quantum-Classical Frontier: Recent Advances in Macroscopic and Mesoscopic Quantum Phenomena*, Eds. J.R. Friedman and S. Han, (unpublished). cond-mat/0004364

17. A. N. Korotkov, cond-mat/0008461

18. T. Christen and M. Büttiker, Phys. Rev. Lett. **77**, 143 (1996)

19. J. Wang, H. Guo, J.-L. Mozos, C. C. Wan, G. Taraschi, and Q. Zheng, Phys. Rev. Lett. **80**, 4277 (1998).

20. B. G. Wang, X. Zhao, J. Wang, et al. Appl. Phys. Lett. **74**, 2887 (1999).

21. M. Büttiker and T. Christen, in *Mesoscopic Electron Transport*, NATO Advanced Study Institute, Series E: Applied Science, edited by L. L. Sohn, L. P. Kouwenhoven and G. Schoen, (Kluwer Academic Publishers, Dordrecht, 1997). Vol. 345. p. 259.

22. M. Büttiker, H. Thomas, and A. Prêtre, Phys. Lett. A **180**, 364 (1993).

23. M. Büttiker, J. Math. Phys., **37**, 4793 (1996).

24. T. Christen and M. Büttiker, Europhys. Lett. **35**, 523 (1996).

25. Z.-s. Ma, J. Wang and H. Guo, Phys. Rev. B **57**, 9108 (1988).

26. E. G. Emberly, G. Kirczenow, cond-mat/0009386

27. M. Büttiker, in "Quantum Mesoscopic Phenomena and Mesoscopic Devices", edited by I. O. Kulik and R. Ellialtioglu, (Kluwer, unpublished). cond-mat/9911188

28. The Lindhard function is $\pi(\mathbf{r}, \mathbf{r}') \equiv \partial\rho(\mathbf{r})/\partial U(\mathbf{r}')$ with ρ the local density and U the local potential. Since we assume a single spatially uniform potential and since we are interested in the total charge on the conductor in the volume (area) Ω we have $\int d^3\mathbf{r} \int d^3\mathbf{r}' \pi(\mathbf{r}, \mathbf{r}') = -e^2 \int d^3\mathbf{r}\nu(\mathbf{r}) = -e^2 D$. Here $\nu(\mathbf{r})$ is the local density of states and D denotes the total integrated density of states of the conductor.

29. V. Gasparian, T. Christen, and M. Büttiker, Phys. Rev. A **54**, 4022 (1996).

30. X. Zhao, J. Phys. Cond. Matter, **12**, 4053 (2000).

31. Ya. M. Blanter, F.W.J. Hekking, and M. Büttiker, Phys. Rev. Lett. **81**, 1925 (1998).

32. P. W. Brouwer and M. Büttiker, Europhys. Lett. **37**, 441-446 (1997).

33. F. T. Smith, Phys. Rev. **118** 349 (1960).

34. G. B. Lesovik, JETP Lett. **49**, 592 (1989).

35. M. Büttiker, Phys. Rev. Lett. **65**, 2901 (1990); Phys. Rev. B **46**, 12485 (1992).

36. M. Büttiker, Phys. Rev. B **41**, 7906 (1990).

37. L. I. Glazman, G. B. Lesovik, D. E. Khmel'nitskii, and R. I. Shekhter, JETP Lett. **48**, 238 (1988).

38. For an ensemble of chaotic cavities the distribution of this conductance derivative is given by P. W. Brouwer, S. A. van Langen, K. M. Frahm, M. Büttiker, and C. W. J. Beenakker, Phys. Rev. Lett. **79**, 914 (1997).

39. A. M. Martin and M. Büttiker, Phys. Rev. Lett. **84**, 3386 (2000).

On the Detection of Quantum Noise and Low-Temperature Dephasing

U. Gavish, Y. Levinson and Y. Imry

Dept. of Condensed-Matter Physics, Weizmann Institute of Science, Rehovot, Israel

Abstract. We discuss the detection of quantum fluctuations in the light of the general van Hove - type relationship between time-dependent correlators and measurable properties. Considering the interaction between the fluctuating electron system and a resonant circuit or a photon mode, we prove that zero point fluctuations (ZPF) *are not observable by a passive detector*, corroborating the results of Lesovik and Loosen. By a passive detector we mean one which is itself effectively in the ground state, and can not transfer energy to the ZPF whose detection is attempted. We find that the ZPF can, on the other hand, *be observed through the absorption spectrum, via the deexcitation of an active detector*. We also make the connection between these statements and the recent discussion of whether decoherence can be caused by the ZPF. We derive a useful general formula for the dephasing rate and use it along with the detailed-balance condition, to prove that the dephasing rate vanishes at the $T \to 0$ limit. The distinction is made between decoherence via making a real excitation in the environment and effects due to its polarization by *virtual* excitations.

1 Introduction and Motivation

Imagine a small isolated normal conducting loop (henceforth the "antenna") at zero temperature. In the absence of external magnetic fields, the average current around the loop vanishes. However, the instantaneous current in the loop is not zero, due to zero-point fluctuations (ZPF). One can in fact calculate, for example, the nontrivial temporal correlation function of these fluctuations and its Fourier transform (the noise power spectrum). The question is now how to detect these ZPF. Naively, one might expect that a very sensitive detector of electromagnetic fields, if brought close enough to the loop, will register these fluctuations. This is in fact true for classical noise, but needs an examination in the quantum realm. Even if this experiment turns out to be difficult, such questions show up in other physical situations. For example, it has been demonstrated that this type of detection is effected by bringing a mesoscopic interferometer to the vicinity of the antenna [1] and the time-dependent fields [2] generated by the latter may dephase the interference in the former. Thus, the dephasing of the interference in a nearby circuit can be considered as a detection of the fluctuations in the antenna. This is known to apply to both thermal [3–5] and shot-noise fluctuations. Therefore, the issues of the detection of fluctuations and the dephasing of quantum interference are intimately related [4]. In particular, the question about the detectability of the ZPF is relevant to that of electronic dephasing in the limit of $T \to 0$.

For the particular case where the fluctuations are in the ground state (ZPF), one may question their simple detectability. In fact, we know that a system in the ground state can not radiate or send out any signal that can be absorbed by the detector. We do not consider here just a "virtual" polarization of the latter by the interaction with the ZPF. Such a polarization will disappear when the detector is moved away from the system and it thus does not constitute a true observation. The electromagnetic fields produced are not classical, and it may follow that they will not be detected unless something more aggressive is done, such as sending radiation to the system in order to measure its absorption, or to induce some nonlinear effects that could be detected. It would then follow that a passive detector, which just "looks" at the system, should not see its zero-point fluctuations, and that a more active type of detection is necessary. In section 2 of this paper we are going to treat this question and show that this is indeed the case, and that no passive detection of zero-point fluctuations is possible. Consequences on the physics of the noise correlators [6] will be obtained. We emphasize that the usual physical effects that are associated with the ZPF, such as the Debye-Waller factor at zero temperature, the Casimir force and the Lamb shift are not affected by this discussion. In those cases the detection of the ZPF is not a passive one. For example, in the case of the Debye Waller factor, radiation is scattered off the system.

In section 3 we are going to discuss the timely issue [7] of electronic dephasing at low temperatures in disordered conductors, using the physical insights gained in the previous section. We will use very general properties of the low-temperature correlators, within a new and powerful formula for the dephasing rate, to demonstrate the vanishing of the latter in the $T \to 0$ limit. This is true for any system that does not have a macroscopic ground state degenracy. This includes virtually all *realistic physical* systems known.

2   Noise Correlators in Mesoscopic Systems

2.1   General Properties of Correlators, Including the Quantum Regime

Let us consider the temporal correlation function of some observable in a general, possibly many-body, and not necessarily an equilibrium, system. While our discussion is completely general and may be relevant to, for example, atomic and optical [8] systems, we will consider here a mesoscopic system, for definiteness. Examples of observables of major interest are the particle density $\hat{n}$ and the current in some direction, $\hat{j}$. In this section, we shall consider the latter. The relevant correlation functions are usually those of a certain Fourier component, $\hat{j}_{\mathbf{q}}$, of the current density, with its conjugate $\hat{j}_{-\mathbf{q}}$. For many real situations it suffices to consider the $\mathbf{q} = 0$ component of $\hat{j}$, which is what we shall do. This corresponds to integrating $\hat{j}$ over the cross section (assumed narrow) and over the length of the system [9]. For interactions with EM fields at all practical frequencies it is easily seen that the relevant values of $\mathbf{q}$ are exceedingly small so that taking only the $\mathbf{q} = 0$ limit (dipole approximation) is quite satisfactory.

The current-current correlation function (denoted here by $C(t)$) is:

$$C(t) = \langle \hat{j}(0)\hat{j}(t)\rangle = \sum_i P_i \langle i|\hat{j}(0)\hat{j}(t)|i\rangle, \tag{1}$$

where P_i is the population of the state $|i\rangle$, and $\sum_i P_i = 1$. The Fourier transform of $C(t)$ is $S(\omega) = \frac{1}{2\pi} \int_{-\infty}^{\infty} C(t) \exp(i\omega t) dt$. As is well known ("Wiener-Khintchine Theorem"), $S(\omega)$ is the power spectrum of the current noise, or the van Hove - type [10] dynamic structure factor for the current. $S(\omega)$ can be very generally shown to be given (see ref. [10]) by:

$$S(\omega) = \hbar \sum_{i,f} P_i |\langle f|\hat{j}|i\rangle|^2 \delta(E_i - E_f - \hbar\omega). \tag{2}$$

Here, $|i\rangle$ is the initial state. At zero temperature, only the many-body ground state, $|g\rangle$ appears with $P_g = 1$, and hence $S(\omega) = 0$ for $\omega > 0$.

More generally, if the system is in equilibrium and its Hamiltonian is time-reversal symmetric, $S(\omega)$ satisfies the important detailed-balance equilibrium relationship, which straightforwardly follows from eq. 2:

$$S(\omega) = S(-\omega)e^{-\hbar\omega/k_B T}. \tag{3}$$

Which also shows how S vanishes for $\omega > 0$ when $T \to 0$. As is well known, S yields the cross section for energy transfer of $\hbar\omega$ from the system ([10]). In our case eq. 2 yields the cross section for emission and absorption of EM radiation as obtained from the golden rule, due to the coupling with the latter at small $\hat{q}$ being given by $\mathbf{A} \cdot \int d\mathbf{r} \hat{\mathbf{j}}(\mathbf{r})$ (see ref. [12]). The vanishing of S for $\omega > 0$ as $T \to 0$ simply means that one can *not* get energy from a system which is in its ground state.

$T = 0$ is an extreme case. More generally, at equilibrium we see from eq. 3, that $S(\omega)$ is not symmetric under reversal of the sign of ω except in the classical limit ($kT >> \hbar\omega$). That certainly happens in many nonequilibrium states as well. Since S is real, C satisfies $C(t) = C(-t)^*$. When $S(\omega) = S(-\omega)$, then $C(t)$ is real and satisfies time-reversal symmetry. However, when S is not symmetric under reversal of ω, C does not respect time-reversal symmetry. Moreover, this also means that C has a nonzero imaginary part. This implies that $C(t)$ is not directly measurable as in the classical case (see below), although both its real and imaginary parts are given by Fourier transforming $S(\omega)$, which is measurable by emission/absorption, inelastic scattering, etc. It is necessary to get $S(\omega)$ for the whole of both the $\omega > 0$ and $\omega > 0$ regimes. The imaginary part and the quantum nature of C are evident for $t < \hbar/k_B T$. For $T \to 0$, there is no classical regime for $C(t)$, where it would be purely real.

In classical systems, $C(t)$ is obtained by measuring $j(t)$ and averaging the product $j(t')j(t + t')$ over, say, the initial time t'. The difficulty in doing this for quantum systems is that the measurement at the earlier time (to obtain $\langle j(t)\rangle$) perturbs the system, so that its evolution to the later time may not be as it should. For example, the value of $j(t + t')$ may be affected by the previous measurement.

The fact that $C(t)$ has an imaginary part should be well-known and some of the above discussion appears already in van Hove's paper [10] for the scattering (but the physics of the correlators discussed there holds much more generally). The imaginary part of $C(t)$ is, of course, due to the noncommutativity of $\hat{j}$ at different times [10]. To understand it even better, write the product of $\hat{j}(0)\hat{j}(t)$ as half the sum of the commutator and the anticommutator. The latter gives the classical symmetrized version of S. The former gives the antisymmetric part of S, and it therefore yields the imaginary part of C. This commutator is in fact what determines the Kubo dissipative conductance, as is well known (and also suggested in ref. [10]). One can say that the net dissipation is actually due to the fact that the absorption and the emission are *not* equal, as is implied by the tendency for equilibration to occur (see subsection 2.3 below). So, while it may appear strange that the correlation function of two physical observables is not purely real, there is nothing wrong about it. The product of $\hat{j}$ at different times is not an hermitean operator, therfore its eigenvalues and expectation values can be complex.

In the next section we discuss the observability of the zero-temperature noise [11] in the light of the above analysis. We corroborate and strengthen the results of ref. [6], that the usual symmetrized version of $S(\omega)$ is *not* a proper noise power spectrum in the low temperature, quantum case. In section 3.1 we derive a simple and powerful expression for the dephasing rate in terms of density correlation functions. In section 3.2, the consequences are obtained for the, recently rather widely discussed, problem of electron dephasing in dirty metals as $T \to 0$. In the last section, we summarize our discussion, in the framework of the apparent misunderstandings of these issues in the recent literature.

2.2 The Detectability of $T = 0$ Noise: A Simple Model

In this section we consider the simplest model for the detection of the current noise by looking at, for example, the photons it produces by linear coupling to an electromagnetic (EM) field. Obviously, this includes such coupling to any set of harmonic oscillators, a simple case being a single one, which can be modelled, for example, by an AC resonant circuit [6].

In ref. [6], the two time-reversed current-current correlation functions were examined for an electronic system

$$C_+(t) = \langle j(0)j(t) \rangle = C(t), \quad and \quad C_-(t) = \langle j(t)j(0) \rangle = C(-t); \quad (4)$$

and their Fourier transforms, $S_+(\omega)$ and $S_-(\omega) = S_+(-\omega)$, respectively. $S_+(\omega)$ is equal to the van Hove-type dynamic structure factor, $S(\omega) = S_{jj}(\omega)$, considered in section 2.1, for which equation 2 holds. Since $C_-(t)$ and $S_-(\omega)$ are simply given in terms of $C(t)$ and $S(\omega)$, it suffices to consider the latter. A free EM field described by a vector potential $\mathbf{A}(x, t)$ is introduced. $\mathbf{A}$ is expanded in terms of photon creation and annihilation operators. By treating the interaction $\int j.\mathbf{A}$ of the electronic system with the EM field by lowest-order perturbation theory, it is straightforward to see [12] that $S(\omega)$ determines the cross section for energy transfer between the fluctuating currents and the EM field. This means that for $\omega > 0$, $S(\omega)$ gives the emission of a photon with a frequency ω into the vacuum, zero photon number, state and $S(-\omega)$ gives the absorption of a given single photon by the electrons. When the electrons are in equilibrium at $T = 0$, the former vanishes, as it should. Regarding the photons emitted into the vacuum state as a means to detect the current noise of the system, it follows that the zero-point fluctuations of the latter are not detectable in this fashion! This is consistent with the second law of thermodynamics, which prohibits the transfer of *any amount of energy* away from a $T = 0$ system.

It is customary (see, for example, Ref. [13]) to take a symmetrized version, $S_s(\omega) = [S(\omega) + S(-\omega)]/2$ given by the Fourier transform of $[C(t) + C(-t)]/2$, for the power spectrum of the current noise. This procedure is fine in the classical limit, but it is *not appropriate* for the quantum fluctuation case, when $k_B T << \hbar\omega$. It implies then that a passive detector would see a half of the absorption spectrum, and it thus misses the quantum effects on the information sent from the system to the detector. This includes, for example, the conclusion that the zero-point fluctuations are detectable by a simple passive detector, which is obviously unacceptable, as discussed above. In the quantum limit, it is the *unsymmetrized* function $S(\omega)$ [6], which corresponds to straightforward physical measurements, which is a *proper noise power spectrum*. Obviously, it is also possible to use the time-reversed version, $S_-(\omega) = S_+(-\omega) = S(-\omega)$, instead of $S(\omega)$, but *not* the symmetrized [13] version $S_s(\omega)$. This problem was considered by Lesovik and Loosen [6]. Our conclusions agree with theirs (see however [14]) but our treatment is more straightforward and free from a mathematical convergence difficulty which is properly acknowledged in their paper. We reemphasize that the asymmetry of $S(\omega)$ at equilibrium, which is dictated by the detailed-balance condition, implies that the time correlator, $C(t)$, has an imaginary part, as discussed in section 2.1.

We obviously do not claim that the zero-point fluctuations (ZPF) are not detectable in general, only that they do not send a signal to a simple *passive* harmonic oscillator detector. It is well-known that the ZPF appear in various other physical effects such as the Lamb shift, the Debye-Waller exponent and the Casimir force. How they influence a linear amplifier is considered for example in ref. [15]. They may well be detectable by various nonlinear phenomena but *not* by the simple direct passive measurement considered above. Furthermore, we are going to demonstrate in the next subsection that augmenting the model by introducing an arbitrary number of photons in the EM field, does produce interesting results, but does not make the ZPF directly detectable by emitting photons. The way the system *absorbs* photons from the field can however be regarded as an observation of the ZPF. This is one of the main interesting results of our discussion. We emphasize that the statement that $S(\omega)$ is the proper noise power spectrum is valid for an arbitrary state (used to perform the averaging in eq.1) of the electronic system. This includes nonequilibrium states, for example current-carrying ones, where shot-noise [16] is relevant. The equilibrium assumption is necessary only if one would like to have the detailed balance condition (Eq. 3) for the correlators.

2.3   Generalization to an arbitrary state of the EM field

When the number of photons in the EM field at frequency ω is $\mathcal{N}(\omega)$, the emission and absorption of photons by the system are modified by the well-known "enhanced emission and absorption" factors $\mathcal{N}(\omega) + 1$ and $\mathcal{N}(\omega)$, respectively. This makes the net energy flow, $I(\omega)$, from the electrons to the field, be given by (we again take $\omega > 0$)

$$I(\omega) \sim [\mathcal{N}(\omega) + 1]S(\omega) - \mathcal{N}(\omega)S(-\omega) \tag{5}$$

$$= S(\omega) + \mathcal{N}(\omega)[S(\omega) - S(-\omega)]. \tag{6}$$

In the particular case where the electronic system is in equilibrium at a temperature T whose inverse is given by $\beta = 1/(k_B T)$, using the detailed-balance condition (Eq. 3) and denoting the equilibrium $\mathcal{N}(\omega)$ at the temperature T by $\mathcal{N}_T(\omega) = 1/[\exp(\beta\hbar\omega) - 1]$, we find:

$$I(\omega) \sim S(\omega)[1 - \frac{\mathcal{N}(\omega)}{\mathcal{N}_T(\omega)}], \tag{7}$$

in agreement with ref[6].

If one assumes, further, that the EM field is at equilibrium at temperature T_0, not necessarily equal to T, one obtains:

$$I(\omega) \sim S(\omega)[1 - \frac{\mathcal{N}_{T_0}(\omega)}{\mathcal{N}_T(\omega)}] = S(\omega)\mathcal{N}_{T_0}(\omega)[\exp(\beta_0\omega) - \exp(\beta\omega)]. \tag{8}$$

It is worthwhile to check that the net enegy flow is from the hotter to the colder system. Such a flow is a way to detect the fluctuations of the system. One may say that at $T = 0$ the zero point noise does not send anything to the detector. In a sense the zero point noise can be observed by *taking* energy from a detector having a nonzero $N(\omega)$. That means that an energetic detector may work at $T = 0$ by *exciting the system*. It is seen from eq.2 that the noise spectrum contains information on the excited states. At $T = 0$ the current correlator is proportional to the Fourier transform of the *absorption* spectrum (i.e. $C(t) = \int_{-\infty}^{0} S(\omega) \exp(-i\omega t) d\omega$). This seems to be a most straightforward way to experimentally obtain the zero-point noise correlator.

3   Dephasing in Mesoscopic Conductors

3.1   A general and useful expression for the dephasing rate

The general picture of dephasing of quantum interference ("decoherence") is by now well understood as due to the coupling of the measured, interfering, particle/wave with other degrees of freedom ("the environment"). The interference is due to at least two paths going through the environmemnt and combining *after* the interaction with the latter has been switched off. Dephasing between two paths occurs when one of them leaves a "which path" trace in the environment by flipping it to a state orthogonal to the one in which the other path leaves it. Alternatively, the dynamic fluctuations of the environment can cause an uncertainty in the relative phase of the two waves which, once large enough, will eliminate the interference. These two seemingly different descriptions can be proven [3] to be precisely equivalent. This equivalence is due to the interaction being the same between the wave and the environment and vice versa (as in Newton's third law), and to a fundamental relationship between the fluctuations of any system and its excitability. It is clear from Eq. 2 that these properties are related to each other. If the system (the environment, in our case), is at equilibrium, this relation is just the fluctuation-dissipation theorem (FDT).

In order to develop a useful general formulation for the dephasing by the coupling to an environment, we start with a test particle whose radius vector is $\hat{x}$, moving in an arbitrary static field. The properties of its motion, for example diffusion in the presence of defects, are assumed to be known. The environment consists of an arbitrary system, whose particles have a set of coordinates denoted collectively by $\hat{r}_s$, and which may comprise nontrivial interactions. Switching off the interaction *between the particle and the environment*, we denote the exact eigenenergies and eigenstates of the former by $E_{p,j}$ and $|j\rangle$, and those for the latter by $E_{s,n}$ and $|n\rangle$, respectively. We now treat the interaction between the test particle and the particles of the environment perturbatively. They are taken to interact via the potential

$$V = \sum_s V(x - r_s) = (2\pi)^{-3} \sum_s \int d\mathbf{q} V_q e^{i q \cdot (x - r_s)}, \qquad (9)$$

where the Fourier transform of the two-body potential $V(r)$ is V_q. Our task is to derive an expression for the dephasing rate of the particle, in terms of the known properties (specifically, the linear response or the dynamic correlation functions) of both the particle and the environment before their mutual interaction has been switched on, to lowest order in the interactions between them. To motivate the later derivation, we first present a "quick and dirty" argument. We start with Eq. 3.28 of Ref.[5], written (using the FDT, ibid. Eq. 3.27) in terms of the dynamic structure factor of the environment $S_s(q,\omega)$:

$$P = \frac{1}{\hbar^2 (2\pi)^3 Vol} \int_0^{T_0} dt \int_0^{T_0} dt' \int d^3 q \int_{-\infty}^{\infty} d\omega |V_q|^2 S_s(q,\omega) e^{i\mathbf{q}\cdot(\mathbf{x}(t)-\mathbf{x}(t'))-i\omega(t-t')}$$

$$(10)$$

P is the probability for the path to have excited the environment during the time τ_0. Remembering that $e^{i\mathbf{q}\cdot\hat{x}(t)}$ is the density fluctuation operator $\hat{n}_q(t)$ of the particle, we consider the sum over classical paths of $e^{i\mathbf{q}\cdot(\mathbf{x}(t)-\mathbf{x}(t'))}$. We argue that by an ergodic-type hypothesis, this sum should be a reasonable representative for the average of the same quantity, $\langle \hat{n}_q(t)\hat{n}_{-q}(t') \rangle$, which is the Fourier transform of the dynamic structure factor, $S_p(q, \omega)$ of the particle. We now convert the two time integrals to two integrals over the average and the relative times. Assuming, further, that the time τ_0 is much longer than all relevant correlation times, makes the integration over the relative times into a Fourier transform and the integral over the average time into a factor τ_0. Thus, the average dephasing probability is linear in time, as long as it is sufficiently small, and the average dephasing rate is given by:

$$1/\tau_\phi = \frac{1}{\hbar^2(2\pi)^3 Vol} \int d\mathbf{q} \int_{-\infty}^{\infty} d\omega |V_q|^2 S_p(-q, -\omega) S_s(q, \omega). \tag{11}$$

It is easy to generalize this to the case where the environment correlators have a slower, for example a power-law, decay. The physical meaning of Eq.11, is simply that the rate of creating *any* excitation in the environment is given by a summation over all the $(\mathbf{q}, \omega)$ channels of exchange between the electron and the environment.

For a more systematic, but nevertheless very simple, derivation of the basic equation 11, we start the system with a direct product state of the particle and the environment, $|im\rangle \equiv |i\rangle \otimes |m\rangle$ and evaluate via the golden rule the rate of transitions into *all different* possible states, $|jn\rangle$. In other words, at any later time, t, the total system's state evolves into

$$|\Psi(t)\rangle = A[|im\rangle + \sum_{j,n} \alpha_{jn}(t)|jn\rangle] \tag{12}$$

where A is a normalization factor. The transition probability is $|A|^2 \sum_{j,n} |\alpha_{jn}(t)|^2$. the transition rate out of the initial state is well-known to be given at all times larger than microscopic by (see, for example, [12,17]):

$$1/\tau_{out} = \frac{2\pi}{\hbar} \sum_{j,n} |\langle im|V|jn\rangle|^2 \int_{-\infty}^{\infty} d(\hbar\omega)\delta(E_{p,j} - E_{p,i} - \hbar\omega)\delta(E_{s,n} - E_{s,m} + \hbar\omega), \tag{13}$$

where the last integral represents the (joint) density of final states, $|jn\rangle$, having the same energy as the initial one, $|im\rangle$. The matrix elements in the last equation are easily evaluated using Eq. 9,

$$\langle im|V|jn\rangle = (2\pi)^{-3} \sum_s \int d\mathbf{q} V_q \langle i|e^{i\mathbf{q}\cdot\mathbf{r}_p}|j\rangle \langle m|e^{-i\mathbf{q}\cdot\mathbf{r}_s}|n\rangle, \tag{14}$$

The absolute value squared of this matrix element consists of "diagonal" ($\mathbf{q} = \mathbf{q}'$), positive and "nondiagonal" ($\mathbf{q} \neq \mathbf{q}'$) terms whose phases are random. An important step is now to average the result over, for example, the impurity ensemble. This will eliminate all the nondiagonal terms. We now introduce a thermal averaging over the initial state, $|im\rangle$, via a summation over i and m, with the factorized weight of that initial state, $P_i P_m$, in obvious notation. It is immediately recognized that the integral in Eq. 13 contains the product of the van Hove dynamic structure factors of the particle and the environment. As a result, we obtain Eq. 11 for the rate $1/\tau_{out}$. We emphasize that this result is exact within the golden-rule formulation which, as is well-known, is much better than just low-order perturbation theory (see, for example, [17]).

Using fashionable terminlogy, $1/\tau_{out}$ is the rate at which the particle gets "entangled" with the environment.

In most situations, as extensively explained in Refs [4,5], $1/\tau_{out}$ is identical with the dephasing rate $1/\tau_\phi$. Important exceptions, having to do with the infrared behavior of the integral in Eq.(11), occur at lower dimensions and their treatment, with the appropriate cutoff [4], is well-known and will be briefly discussed below.

For a diffusing electron, the dynamic structure factor is given, for $\hbar\omega \ll k_B T$, by a Lorentzian with width Dq^2. The low-temperature case will be discussed later. Using for $S_s(q,\omega)$ the dynamic structure factor of the electron gas, given to leading order (see Eqs. 3.28 and 3.44 of Ref.[5] and [3,18]) by $\frac{\hbar q^2 \omega Vol}{2(2\pi)^3 e^2 \sigma}$, allows for an extremely simple calculation of the dephasing rate by electron-electron interactions. Evaluating the integrals in Eq. 11 at dimensions above 2, is straightforward, and it produces identical results to those of Refs. [3,4]. For 2D and 1D (thin films and wires) the small q subtleties are again the same as discussed in these references. The integrations over the appropriate components of $\mathbf{q}$ are replaced by summations and it is found that the remaining integrations are infrared (small q)-divergent. This divergence is again cured by a cutoff which can be regarded as due to an effective finite time τ_ϕ and length scale L_ϕ, and whose physical meaning is exactly that low q excitations can not distinguish paths that are separated in space by less than $\sim 1/q$.

3.2 Dephasing when $T \to 0$

Recently, Mohanty et al[7] have published extensive experimental data indicating that contrary to general theoretical expectations [3-5], the dephasing rate in films and wires does not vanish as $T \to 0$. Serious precautions[19] were taken to eliminate experimental artifacts. It was speculated that such a saturation of the dephasing rate when $T \to 0$, might follow from interactions with the zero point motion of the environment. These speculations have received apparent support from calculations in ref.[20]. However, the latter were severely criticized in refs.[21,22] and were in disagreement with experiments in ref.[23]. In fact, it is clear that since dephasing must be associated with a change of state of the environment [4], it cannot happen as $T \to 0$ [26]. In that limit neither the electron nor the environment has any energy to exchange. Below, we review the proof [24,27] of this qualitative statement, based on Eq.11 and the detailed balance condition, eq. 3. While proving unequivocally that zero point motion does not dephase, the proof does show what *further* physical assumptions can in fact produce an apparently finite dephasing rate when T is very small, but not in the strict $T \to 0$ limit.

It is useful to apply the expression 11 ([24,25,27]) for the dephasing rate, with the very general detailed-balance equilibrium relationship, eq. 3

$$S(q,\omega) = S(-q,-\omega)e^{-\hbar\omega/k_BT}, \tag{15}$$

to either $S_p(q,\omega)$ or $S_{env}(-q,-\omega)$. It is immediately seen that the integrand of eq.11 is a product of two factors one of which vanishes for $\omega > 0$ and the other for $\omega < 0$, as $T \to 0$. Thus the integral and the dephasing rate vanish in general when $T \to 0$. The dephasing of a conduction electron at the Fermi level consitutes a very sensitive passive detector for the fluctuations of the environment. However, it detects nothing when the environment is in its ground state as well. One may note, however, that if $S_{env}(-q,-\omega)$ has an approximate delta-function peak at small ω, due to an abundance of low-energy excitations, relatively strong dephasing will follow at the correspondingly low temperatures. A particular model for those, invoking defect dynamics, was suggested in ref. [27].

Thus, the "standard model" of disordered metals (in which the defects are strictly frozen) gives an infinite τ_ϕ at $T = 0$. But there may be other physical ingredients that can make τ_ϕ finite at very low temperatures (but *not at the* $T \to 0$ *limit*), *without* contradicting any basic law of physics. The model considered in ref. [27] is a particular example and its requirements may or may not be satisfied in the real samples. But, other models with similar dynamics might exist as well. Isolated magnetic impurities above the Kondo temperature, might serve as another example. All such models are not "universal" and their properties depend on the purity and metallurgy of a given sample. This is consistent with the fact [28] that silver has no "saturation" while similarly prepared copper and gold do (but on different, nonuniversal levels) [28]. In fact, this also disproves the universal nature of the presumed saturation, which depends only on the density of states, mean free path and thicknesses of the samples [20]. The finite measurement current and lack of full electronic equilibrization may also play an important role [31]. We reemphasize that the increased dephasing at low but finite temperatures, does *not* imply dephasing by zero-point fluctuations, which has been repeatedly, and wrongly, claimed in the literature. The failure of the semiclassical approximation used in these considerations was clarified in ref.[24].

4   Conclusions

There is no doubt that the ZPF exist and are measurable, using the right experimental arrangement (in fact, we have shown that the ordinary absorption spectrum provides such a measurement). However, Perhaps the main point of this paper is that one should *not* think naively that the EM fields produced by the fluctuating charges and currents are directly measurable by a passive detector. This is what one may be led to believe by reading, for example, refs. [20] and [29]. In quantum Physics, one does not directly measure the time dependence of operators, only expectation values are accessible. If the Hamiltonian is time independent then the expectation value of the current in any stationary state, including the ground state, is time independent. The current dynamic correlators (as in eq. 1) are generally nonzero and time-dependent also in the ground state. We proved here, however, that they are not measurable through signals that they send to passive detectors, but can very well be measured for example via the nonzero absorptive part of their Fourier transform $S(\omega)$. Thus, care is needed in using too vividly the picture [29] of flows of matter and energy between subsystems when the total closed system is in the ground state.

The reduction of the $T = 0$ persistent current in a mesoscopic ring with a resonant transmitter reported in ref. [29] is a true effect. It is due, however, to the reduction of the tunneling amplitude between the resonant level and the ring [30], by a reversible polarization of the environment. Only *virtual* excitations of the latter are produced but no real excitation, needed [4] to cause "decoherence". Whether the above persistent current reduction is a true "decoherence" effect was not clearly spelled out in ref. [29] (although it was repeatedly hinted at). We would like to make here the unambiguous statement that this reduction has nothing to do with $T = 0$ "decoherence". Strict dephasing is *only* caused by a real change in the state of the environment. This is to be contrasted with a reversible polarization of the latter. Modelling it, for example, by a set of independent harmonic oscillators linearly coupled to the electron coordinate, it is found that, for example, the tunneling amplitude between two given quantum states is in general reduced. The oscillators are polarized differently when the electron is in one state or in the other. The tunneling amplitude between these two states is reduced by the the product of the overlaps between the two shifted states of each oscillator. Under some conditions, this total overlap can even vanish [30]. However, only *virtual* excitations of the environment oscillators are produced, but no real excitation, needed [4] to cause "decoherence". Such a reversible polarization of the environment by two partial waves is different from a real decoherence. It occurs only as long as the environment interacts with the electron. The polarization disappears if the coupling of the electron with the environment is switched off. This happens, for example, when the electron leaves the medium and is far enough from it during the observation of the interference. That is the physical situation appropriate also for typical transport measurements, which we have considered here. An electron wave whose energy is just at the Fermi level will emerge after spending an arbitrarily long time in an environment which is in a non-degenerate ground state, without creating any real excitation in it. This wave will therefore retain full coherence with another partial wave which has not interacted at all with the environment. This is very different from the case where a real excitation of the environment is produced by one of the partial waves. The latter will imply that the two partial waves will leave the environment in two orthogonal states. This orthogonality lives forever after the coupling has been switched off [4] and it causes a real dephasing of the electronic interference. The overlap reduction by virtual excitations is just a renormalization [26] of the parameters and it is similar to the zero-temperature

Debye-Waller factor in coherent scattering, which has nothing to do with "$T = 0$ decoherence".

5 Acknowledgements

This research was supported by grants from the German-Israeli Foundation (GIF), the Israel Science Foundation, Jerusalem and from the Israeli Ministry of Science and the French Ministry of Research and Technology. Instructive discussions with Y. Aharonov, D. Cohen, S. Haroche, J-M Raimond, A. Stern , D.E. Khmelnitskii, the late R. Landauer, Z. Ovadyahu, A. Schwimmer, B.Z. Spivak, and P. Wölfle are gratefully acknowledged. Some of this work was done when YI was visiting the Ecole Normale Superieure in Paris. He thanks S. Haroche for his hospitality and the Chaires Internationales Blaise Pascal for support.

References

1. E. Buks, R. Schuster, M. Heiblum, D. Mahalu and V. Umansky, Nature **391**, 871 (1998).
2. Y. Levinson, Europhys. Lett. **39**, 299 (1997).
3. B.L. Altshuler, A.G. Aronov, and D.E. Khmelnitskii, J. Phys. **C15**, 7367 (1982) .
4. A. Stern, Y. Aharonov, and Y. Imry, Phys. Rev. **A41**, 3436 (1990) ; and in G. Kramer, ed. *Quantum Coherence in Mesoscopic Systems*, NATO ASI Series no. 254, Plenum., p. 99 (1991).
5. Y. Imry, *Introduction to Mesoscopic Physics*, Oxford Unversity Press (1997).
6. G.B. Lesovik and R. Loosen, JETP Lett., **65**, 295 (1997); see also: G.B. Lesovik and L. S. Levitov, Phys. Rev. Lett **72**, 538 (1994).
7. P. Mohanty, E.M. Jariwala and R.A. Webb, Phys. Rev. Lett. **78**, 3366 (1997); P. Mohanty and R.A. Webb, Phys. Rev. **B55**, 13452 (1997).
8. For the Quantum Optics context, see for example, S. Haroche, Physics Today **51** (7), 35 (1998).
9. For times longer than the time needed for the charged particle to cross the system, the current is approximately conserved along its length, so that the latter integration is immaterial. The system is taken to be short enough, so that the interesting quantum effects will show up at the above time scales. In usual mesoscopic systems, this is seen to necessitate temperatures smaller than the relevant Thouless energy.
10. L. van Hove, Phys. Rev **95**, 249 (1954). In this classic paper, the dynamic correlation function for the density $\langle \hat{n}_q(0)\hat{n}_{-q}(t) \rangle$, was introduced, it was shown that its time Fourier transform, $S(q,\omega)$, yields the inelastic Born scattering cross section from the system, the properties of the latter were analyzed and the connection to the dissipative response indicated.
11. U. Gavish, Y. Levinson and Y. Imry, Phys. Rev. **62**, R10637 (2000).
12. G. Baym, *Lectures on Quantum Mechanics*, p. 271-276, Addison-Wesley (1993).
13. L. D. Landau and E. M. Lifschitz, *Statistical Physics*, Pergamon (1958).
14. G.B. Lesovik, Phys. Usp., **41**(2), 145 (1998)
15. J. R. Tucker and M. J. Feldman, Revs. Mod. Phys. **57**, 1055 (1985).

16. V.A. Khlus, JETP **66**, 1243 (1987) ; G.B. Lesovik, JETP Lett., **49**, 592 (1989); Th. Martin and R. Landauer, Phys. Rev. **B45**, 1742 (1992); M. Büttiker, Phys. Rev. **B46**, 12485, (1992).

17. E. Merzbacher, *Quantum Mechanics*, John Wiley and Sons (1970).

18. Y. Imry, Y. Gefen, and D. J. Bergman, Phys. Rev. **B26**, 3436 (1982).

19. R.A. Webb, P. Mohanty and E.M. Jariwala in *Quantum Coherence and Decoherence*, proceedings of ISQM, Tokyo (1998), K. Fujikawa and Y. A. Ono, eds. North Holland, Amsterdam (2000).

20. D.S. Golubev and A.D. Zaikin, Phys. Rev. Lett., **81**, 1074 (1998); Phys. Rev. **B59**, 9195 (1999); Phys. Rev. Lett., **82**, 3191 (1999).

21. I. L. Aleiner, B. L. Altshuler, M. E. Gershenson, Waves in Random Media **9**, 201 (1999); Phys. Rev. Lett., **82**, 3190 (1999).

22. B. L. Altshuler, M. E. Gershenson, I. L. Aleiner, Physica **A3**, 58 (1998).

23. Yu. B. Khavin, M. E. Gershenson and A. L. Bogdanov, Phys. Rev. Lett., **81**, 1066 (1998), Phys. Rev. **B58**, 8009 (1998); M. E. Gershenson et al., Sov. Phys. Uspekhi **41** (2), 186 (1998).

24. D. Cohen and Y. Imry, Phys. Rev. **B59**, 11143 (1999).

25. Y. Imry, as in ref. [19].

26. Interestingly, this issue appears to have been settled already in 1988. See, for example, J. Rammer, A. L. Shelankov and A. Schmid, Phys. Rev. Lett. **60**, 1985 (1988).

27. Y. Imry, H. Fukuyama and P. Schwab, Europhys. Lett, **47**, 608 (1999).

28. A. B. Gougam, F. Pierre, H. Pothier, D. Esteve and N. O. Birge, J. Low Temp. Phys. **118**, 447 (2000); see also: F. Pierre, H. Pothier, D. Esteve and M. H. Devoret, ibid.; F. Pierre, Ph.D. thesis, Universite Paris VI (2000), unpublished.

29. P. Cedraschi, V. V. Ponomarenko and M. Büttiker, Phys. Rev. Lett. **84**, 346 (2000).

30. A. J. Leggett et al., Rev. Mod. Phys. **59**, 1 (1987); A. Stern, Ph.D. Thesis, Tel-Aviv University (1990), unpublished.

31. Z. Ovadyahu, unpublished (2000).

Part V

Spintronics and Magnetism

Spintronics: Spin Electronics and Optoelectronics in Semiconductors

Michael Oestreich[1,2], Jens Hübner[1], and Daniel Hägele[1]

[1] Philipps-Universität Marburg, Fachbereich Physik und Wissenschaftliches Zentrum für Materialwissenschaften, Group of Wolfgang W. Rühle, Renthof 5, D-35032 Marburg, Germany
[2] Universität Hannover, Institut für Festkörperphysik, Abteilung Nanostrukturen, Appelstraße 2, D-30167 Hannover, Germany

Abstract. Although an electron has electric charge and spin, today's semiconductor devices are restricted to the precise control of the charge only. Taking additional advantage of the two possible electron spin orientations – spin up and spin down – might revolutionize electronics. What will be the advantages of this new technique, how far is it developed, where are the problems, and when can we buy the first spin electronic computers?

1 Introduction

The control of electrical charge has influenced nearly all aspects of life during the last hundred years. Now at the beginning of a new millenium another property of the electron might rush forward and change today's semiconductor industry, the spin moment (see box on the next page for an explanation of spin). In metals, the electron spin is already employed in some simple devices: reader heads of magnetic hard disk drives are the first devices with major economic impact of magnetoelectronic (the spin electronics in metals). Here the use of the giant magnetic resistant (GMR) [1,2] effect empowers the reading of extremely high density computer disks. Another application of the GMR effect is being developed at present, as Honeywell demonstrated a giant–magnetoresistance random access memory, known as MRAM, which might push forward very soon in the 100 billion semiconductor DRAM market. MRAMs are simple nanoscale computer memories, which have in comparison to DRAMs the advantage to be non-volatile, are about a million times faster compared to traditional magnetic hard disks, and rapidly approach the speed and density of semiconductor DRAMs.[3] More futuristic applications of spin electronics are, e.g., nonvolatile reprogrammable logic gates which could be fabricated by magnetoelectronic elements. Employing these magnetoelectronic elements, microprocessor chips could reconfigure themselves in midcalculation at nanosecond speed just by reversing the magnetization of some elements in order to most efficiently address the next part of the calculation.[4]

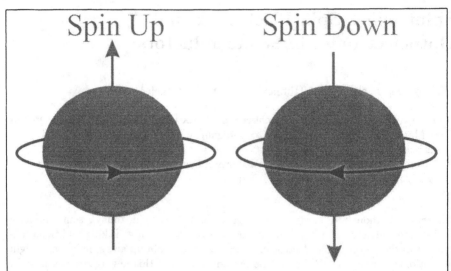

Electrons are fermions with spin 1/2, i.e., graphically they are small balls who spin with a fixed frequency either clockwise or counterclockwise around their rotational axis, but actually this helpful image of the spin is inaccurate. The spin is a quantum mechanical property and has in any given direction only two possible orientations, "Up" or "Down", which are labeled by their quantum numbers +1/2 and -1/2, respectively. The limitation to only two orientations cannot be explained by classical physics but exactly this limitation makes the electron spin a good candidate for the representation of exactly one bit, 0 or 1. Connected with the spin s is the magnetic moment $\mu_B = -g_e Bs$, where g_e is the electron Lande g factor and B is Bohr's magneton. A magnetic field perpendicular to the spin moment forces the spin to precess around the magnetic field. The precession frequency is proportional to the applied magnetic field and the electron Lande g factor. While g_e equals 2.0023 for free electrons in vacuum, g_e in semiconductors differs considerably and is either positive, negative, or even equal to zero. The value of g_e can be engineered by the semiconductor material and semiconductor structure.[5] The absolute value of g_e can be extremely large and is, e.g., about -50 in InSb and up to several hundreds in diluted magnetic semiconductors. Such large g_e might be interesting for sensitive magnetic field sensors. On the other hand, g_e can be equal to zero which might be useful to store information since the spin orientation is insensitive to external magnetic fields in that case.

All these magnetoelectronic devices are very promising but an epoch-making spin electronics must be compatible to the existing semiconductor platform, where physicists can easily engineer the electronic properties, combine conventional and spin electronics on one chip, and build optoelectronic spin devices. Of course one may ask if spintronics, the spin electronics in semiconductors, and especially spin optoelectronics is only an enticing dream but first experiments strongly encourage intensified research. As an example for the first step towards spin optoelectronics, figure 1 shows the ultrafast optical switching of a vertical-cavity surface-emitting laser (VCSEL) by modulation of the electron spin orientation only. The VCSEL consists of (GaIn)As quantum wells as active

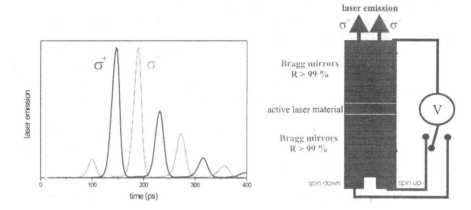

Fig. 1. (left) Ultrafast optical modulation of a VCSEL by modulation of the electron spin orientation only. The VCSEL is pumped by circularly polarized laser pulses along the growth direction exciting electrons with a preferential spin orientation in the same direction. The electron spins precess in a transverse magnetic field of 2 T with a frequency of 22 GHz. As the electron spins oscillate between spin up and down the laser output alternates between right and left circularly polarized light. (right) Suggestion for a simple electro-optical spin device. Spin-polarized electrons are electrically injected over two distinct spin-polarizing contacts. The VCSEL emission is modulated by changing the electrical injection between spin up and/or spin down electrons without changing the total carrier density.

laser material sandwiched between highly reflecting dielectric mirrors made by alternating layers of GaAs and AlAs. Spinpolarized electrons are injected into the active laser material by using a circularly polarized pump laser pulse tuned above the GaAs bandgap. The optically excited electrons have a fixed initial spin orientation in the GaAs due to the optical selection rules and fall into the (GaIn)As quantum wells without losing their spin orientation. The electron spins precess around an external transverse magnetic field periodically changing the population of spin up and spin down states. Since the electron spin orientation determines the polarization of the emitted light, the VCSEL periodically emits right and left circularly polarized light. In between the laser switches off since the population of the carriers is divided between spin up and down states and the laser falls below threshold. In this manner S. Hallstein et al. showed spin controlled laser emission with a repetition frequency of 22 GHz - at a field of 2 T - and a large peak to valley ratio of more than 96%.[6] Even higher frequencies in the terahertz regime [7] should be feasible in semiconductors with a larger electron g-factor as, e.g., in InSb or in dilute magnetic semiconductors.

How would a real device look like? The right part of figure 1 shows one very simple suggestion using this spin optoelectronic concept. Spin-polarized electrons are electrically injected by spin-polarizing contacts causing the VCSEL to be modulated at constant current in any given sequence between σ^+ and σ^--emission just by switching between the two different spin-polarizing contacts. Operating a laser with a constant current is attractive since undesirable temperature and carrier relaxation effects are reduced in respect to electrical power on and power off operation and the maximum laser modulation-frequency increases. On the other hand, if also the current is modulated the "spin laser" could carry twice the information compared to a conventional laser at the same repetition rate (σ^+, σ^-, $\sigma^+ + \sigma^-$, or no emission instead of on or off). The modulation of the microcavity laser demonstrates that the control of the entity spin might be interesting for future semiconductor devices. But how far are we away from commercial spintronics? To be honest, spintronics is still in its infancy and faces several basic problems. The first and perhaps most profound and most critical point seems trivial at first glance: the electrical injection of spin-polarized electrons, i.e., all or at least most injected electron spins must point into the same direction. Then, as a second task, the direction of the spin must be preserved as the electrons move in an electric field through the semiconductor. Thirdly, it should be possible to manipulate the spin in a predefined manner by electric and/or magnetic fields. Fourthly, the storage of the spin orientation over a long time range is desirable. And fifth, all this should work at room temperature. While the first two requirements are necessary for nearly all spintronic devices, the later two are the basic conditions for advanced spintronic information processing, and the last is the usual requirement for commercial success.

2   Spin Injection

In magnetoelectronics, the injection of spin-polarized electrons from a ferromagnetic into a nonmagnetic metal and visa versa was a cornerstone for spin dependent electronics.[8] The use of a ferromagnetic metal was straight forward since ferromagnets exhibit a natural electron spin orientation at the Fermi energy. Generalizing this concept to a ferromagnet-semiconductor interface also seemed straight forward, but despite considerable effort by many different groups it remained elusive so far. In 1992 Alvarado and Renaud demonstrated in a scanning tunneling microscope (STM) experiment injection of spin-polarized electrons from a ferromagnetic Ni tip through a vacuum barrier into p-doped (110) bulk GaAs.[9] The Ni tips had an apex radii of 40 to 70 nm and were cleaned in ultrahigh vacuum by Ne-ion bombardment and subsequent heating. Alvarado and Renaud magnetized the tip in situ by an electromagnet and measured the spin polarization of the injected electrons by detecting the polarization of the emitted GaAs photoluminescence. The measured spin polarization of the injected electrons was in this room temperature experiment unbelievable high and reached at low injection energies more than 30 %. Of course, an STM tip is not applicable in spintronic devices but the experiment strongly motivated further experiments concerning spin injection from ferromagnetic metal contacts into semiconductors. At the same time, the experimental results gave a first warning. Alvarado and Renaud found that insufficient cleaning of the tip or contamination of the apix results in a loss of the spin polarization associated with the tip magnetization. "This means that a low number of contaminant atoms, maybe even a single one, can drastically influence the spin polarization of the charge carriers tunneling into the GaAs." About seven years later two groups, P. R. Hammar et al. from the Naval Research Laboratory [10,11] and W. Y. Lee et al. from the University of Cambridge [12,13], reported independently the observation of spin injection at a ferromagnet- semiconductor interface in a ferromagnet-semiconductor-ferromagnet device which was suggested theoretically by Datta and Das [14]. We discuss in the following only the first since both experiments are very similar. Hammar et al. projected the spin-polarized current of a ferromagnet onto the spin split density of states of a high mobility two dimensional electron gas of an InAs quantum well and measured the change of interface resistance when the magnetization orientation of the ferromagnet was reversed. The interface resistance changes were of the order of 1 % at room temperature indicating according to their interpretation an interfacial current polarization of about 20 % at the ferromagnetic Permalloy (80 % Ni, 20 % Fe) contact. However, what seems at a first glance like a proof of electrical spin injection yielded more controversy than clarity.[15,16] The change in resistance is very small and can not only be attributed to spin injection but also to parasitic effects as, e.g., stray-field induced Hall or magnetoresistance effects. To really proof spin injection by transport measurements much higher changes in resistance are necessary but theory shows that there are fundamental difficulties to achieve much higher interface resistance changes in ferromagnet-semiconductor-ferromagnet devices if the degree of spin polarization is not close to 100%.[17] In Permalloy contacts, the degree of spin polarization of the injected carriers can not be higher than 40% [18] since only the d- electrons are aligned at the Fermi energy but not the s-electrons. Ferromagnetic contacts made by materials with 100% spin polarization at the Fermi energy are difficult to fabricate on semiconductors since the processing temperatures of these materials are usually too high for customary semiconductors. The possibility of parasitic effects is not the only reason why these experiments are discussed so controversely. Another reason are optical experiments in the manner of Alvarado and Renaud which are much more sensitive to spin polarization and much less sensitive to parasitic effects. In quantum wells, for example, the circular polarization of the photoluminescence is identical to the spin polarization of the electrons but, so far, optical experiments did not confirm spin injection from ferromagnetic contacts into a semiconductor. What are the reasons that spin injection from a ferromagnetic tip into a semiconductor is feasible but injection from a thin ferromagnetic layer is so difficult? One explanation discussed in literature is related to the grouping of magnetic domains in ferromagnetic layers. Large ferromagnetic electrodes with edge lengths much larger than 1 μm are usually divided in several domains with varying magnetic orientation minimizing the spin polarization of the injected carriers. The requirements of a well defined magnetization can be satisfied by single domain type electrodes where the spins in the ferromagnetic material point into the same direction. However, single domain electrodes are extremely small and generate a strong magnetic stray field in the neighboring semiconductor. The stray field randomizes the spin of the injected electrons and generates at the same time a spatially varying Hall effect in the semiconductor making measurements of the injection of spin-polarized electrons very difficult or even impossible. Meier and Matsuyama from the university in Hamburg suggested a solution to this problem by using only one domain of a specially tailored multidomain structure.[19] The single domain electrical injection should lead to a high degree of spin polarization and the tailored multidomain structure should minimize the effects due to the stray fields. They verified the simulated magnetization of a Permalloy electrode configuration on the semiconductor InAs by magnetic force microscopy but spin injection using this concept still has not been experimentally demonstrated so far. Another often discussed explanation, why spin injection from a ferromagnet into a semiconductor is so difficult, are dead layers at the ferromagnet semiconductor interface. Dead layers are monoatomic layers directly at the interface who have in contrast to the bulk ferromagnet random magnetic orientations and act as very efficient spin scatterer and depolarizer. The group of G. Bayreuther from the university in Regensburg was one of the first who demonstrated the growth of dead layer free ferromagnetic contacts [20] but its application to spin injection was also not successful. Therefore, despite great experimental effort convincing electrical spin injection from a ferromagnetic metal into a diffusive semiconductor has *not* been observed, [21] which is consistent with calculations by G. Schmidt et al..[17]

Are the huge difficulties to inject spin-polarized electrons by ferromagnetic contacts already the early death of spintronics or are ferromagnetic contacts only the wrong approach?

One very promising alternative approach has been suggested and experimentally demonstrated at the university of Marburg: injection of spin-polarized electrons via a spin aligner based on diluted magnetic semiconductors.[22] How does it work? Figure 2 shows the principle idea. Unpolarized electrons are injected from a metal into a diluted magnetic semiconductor incorporating, e.g., Mn. The spins of the injected electrons interact with the $S = 5/2$ spins of the localized $3d^5$ electrons of the Mn ions via the sp-d exchange interaction. Hence, already a small external magnetic field, which can even be supplied by micro magnets, gives rise to a giant effective Zeeman splitting of the conduction band states, and the spins of the unpolarized injected electrons align on a picosecond time-scale antiparallel to the magnetic field. The feasibility of this approach has first been proofed by an all optical method [22] and has recently been confirmed electrically by the group of L. Molenkamp from the university in Würzburg and independently by a collaboration led by Berry Jonker from the Naval Research Laboratory and Athos Petrou from the State University of New York in Buffalo in a light emitting device (LED) structure.[23,24] Such all-semiconductor devices offer several advantages over ferromagnet-semiconductor devices since

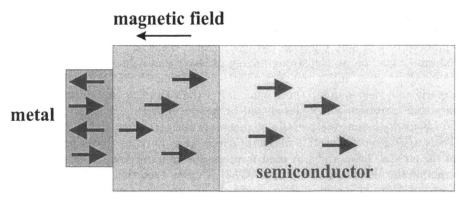

magnetic field

metal

semiconductor

**spin aligner
(diluted magnetic semiconductor)**

Fig. 2. Injection of spin polarized electrons using a diluted magnetic semiconductor as spin aligner.[22] Unpolarized electrons are injected from the metal contact into the spin aligner where the electron spins rapidly align antiparallel to the magnetic field. So far all experiments were done with external magnetic fields but micromagnets as metal contacts should be feasible.

the interfaces, band offsets, and carrier densities can be perfectly controlled and their production can be easily integrated in existing semiconductor and fabrication technology. What is even better, the experimentally demonstrated degree of spin polarization from the diluted magnetic into the nonmagnetic semiconductor is as high as 90%.

Is spin injection thereby solved? The answer for spintronic devices is unfortunately no. The demonstrated spin aligners out of II-VI diluted magnetic semiconductors work well at low temperatures but so far this material system only exhibits its special magnetic properties at temperatures much below liquid nitrogen (i.e., a few Kelvin). To improve the temperature range by increasing the Mn concentration is difficult, because the Mn ions couple antiferromagnetically in this system at high concentrations.[25]

The group of Ohno from the Tohoku University therefore started together with the group from D. Awschalom from the University of California in Santa Barbara (UCSB) a different approach.[26] They use p-GaMnAs (a group III-V semiconductor) as spin aligner to inject positively charged holes rather than electrons. GaMnAs is intrinsically p-doped if the Mn ions are substitutional on the lattice site of the Ga ions and has the huge advantage over II-VI semiconductors to be ferromagnetic even at moderate temperatures. Ferromagnetic semiconductors are superior over paramagnetic semiconductors concerning spin injection since they do not to require an external magnetic field to align the electron spins. So far, a Curie temperature of 110 K has been demonstrated in $Ga_{0.95}Mn_{0.05}As$,[27] but even room temperature ferromagnetism might be feasible if the Mn concentration can be further increased or Mn-doped GaN or ZnO is employed.[28] Unfortunately, the interaction between the Mn ions leading to ferromagnetism is mainly mediated by holes in these semiconductors, and the approach of hole spin injection has one intrinsic disadvantage: hole spin flip is extremely fast due to the strong mixing of heavy and light hole in the valence band. Ohno et al. inject the spin-polarized holes into bulk GaAs where other experiments yield spin flip times much faster than 1 ps.[29] Due to the fast spin flip, spin transport with holes should be nearly impossible in bulk GaAs over non ballistic distances especially at moderate temperatures. Consequently, the Tohoku/UCSB collaboration measures a more than ten times lower polarization of the emitted light, which is used to probe the spin injected carriers, than for example the Würzburg group does. What is more, even this small polarization does not prove hole spin injection and hole spin transport unambiguously since they use a light detection geometry where the polarization of the hole spins is not easily related with the polarization of the emitted light via the optical selection rules. Spin injection for room temperature spintronic devices is therefore at the moment an unsolved problem but probably not an unsolvable problem. As spin aligner, any ferromagnetic semiconductor that injects electrons and works at room temperature would do the job. On the other hand, progress on the physics of ferromagnetic metal contacts for spin injection is fast. Which approach will win this race is unclear at the moment, and it might even be a combination of ferromagnetic metal and diluted magnetic semiconductor.[30]

3   Spin Transport

The second important requirement for spintronic is transport of the spin through the semiconductor. The important question is if the spin information can be spatially transferred in high electric fields over macroscopic distances. We will exclude in the following results concerning ballistic transport [31–33] and focus our attention to the device relevant diffusive transport where the electrons perform many collisions. Daniel Hägele et al. were to our knowledge the first who studied this kind of transport within a semiconductor.[34] They circumvented the problem of electrical spin injection by creating spin-oriented electrons optically with circularly polarized light and studied spin transport in GaAs over a distance of 5 μm in electric fields as high as 6 kV/cm. The degree of spin polarization was detected by an optical marker where the electrons were captured after transport. The degree of polarization of the photoluminescence from the optical marker directly yielded the spin polarization of the electrons after transport. No measurable dephasing was observed. A short time later J. M. Kikkawa and D. D. Awschalom showed spin transport over even longer distances by Faraday-rotation measurements. The Faraday-rotation measurements base upon a special pump-probe technique where a circularly polarized pump pulse creates spin-polarized electrons and the Faraday-rotation of a temporally delayed and spatially varied linearly polarized probe pulse yields the temporal and lateral spin dynamics. They measured spin transport at low electric fields of 16 V/cm over distances exceeding 100 μm. The characteristic exponential decay times of the spin orientation were 29 ns without and 17 ns with electric field, demonstrating that applying a voltage does not necessarily introduce severe spin decoherence. The experiments from the UCSB group also turn up a more subtle effect. Connected with the spin transport is the flow of an electrical current which by itself generates an intrinsic magnetic field. The measurements indicate that this magnetic field can lead to spin dephasing if the current is non-uniform or even filamentary. Therefore in future spintronic devices, the design of current leads could be used to manipulate the spin. On the other hand, the current leads must be very carefully designed to avoid unintentional spin manipulation. Spin transport is not restricted to bulk semiconductors. Sogawa et al. from the NTT Basic Research Laboratories in Kanagawa showed for example electron-spin transport in p-type rectangular GaAs/AlAs quantum wires with small (10 nm) lateral size over distances of 10 μm.[36] Such an effective spin transport in quantum wires is not a direct matter of course - as it might look like - due to the high conduction band spin splitting at finite carrier momentum.

Is spin transport in semiconductors perhaps restricted to low temperatures? All the above measurements were done at temperatures of a few Kelvin, but what happens at room temperature? Are there as many contrarieties effective as in the spin injection experiments? Measurements of the electron spin dephasing without electric field yield extremely long spin dephasing times at low temperatures. Kuzma et al. measure a longitudinal relaxation time of an electron gas magnetization of more than 100 ns in GaAs quantum wells in the quantum Hall regime at 300 mK. At more moderate temperatures of 5 K, spin dephasing times of 100 ns were measured in n-doped bulk GaAs, decreasing with increasing temperature down to 0.5 ns at 150 K.[37] At room temperature, Takeuchi et al. observe in undoped InGaAs quantum wells electron spin relaxation times of less than 6 ps - that is more than four orders of magnitude faster than at 5 K in n-doped GaAs.[38] Although, such very fast spin relaxation at room temperature can be used for fast all-optical switching devices[39], they are unsuitable for spintronics.

What are the reasons for fast spin dephasing? The causes of spin relaxation in semiconductors are still not completely understood but several mechanisms are clearly identified. The importance of these mechanisms depends on a number of factors, e.g., the semiconductor material, carrier type and density, momentum relaxation times, and temperature. One important spin relaxation mechanism is the Bir-Aronov-Pikus mechanism caused by exchange and annihilation interaction between electrons and holes. This mechanism is especially efficient in p-doped semiconductors at low temperatures.[40,41] Another spin relaxation mechanism is the Elliot-Yafet mechanism which results from spin orbit scattering during collisions with phonons or impurities. This mechanism is important at low and moderate temperatures but less efficient at high temperatures. At high temperatures, a third mechanism usually dominates in III-V semiconductors, the D'yakonov-Perel (DP) mechanism.[42] In the DP theory, the driving force for spin relaxation is the spin splitting of the conduction band via spin-orbit coupling due to the lack of inversion symmetry in III-V compounds. The DP mechanism is, for example, responsible for the fast spin dephasing in bulk GaAs at elevated temperatures. Further decrease of the symmetry and increase of the momentum due to quantum confinement increases the DP spin relaxation both in (100) oriented QWs and in quantum wires [43]. One requirement for room temperature spin transport in III-V semiconductors is therefore the suppression of the DP mechanism. This is possible since the DP interaction depends on the semiconductor material and the direction of electron momentum and spin in the host crystal. Ohno et al. demonstrated the suppression of the DP mechanism by choosing a quantum well with special crystal axises, an undoped (110) oriented GaAs QW, and observed even a decrease of nearly one order of magnitude of the spin relaxation time with increasing temperature reaching spin relaxation times as long as 2 ns at room temperature.[44] This is very much longer than in (100) oriented QWs. An even stronger increase of nearly two orders of magnitude of the spin relaxation time with temperature is observed in ZnSe and ZnCdSe QWs.[45,46] The slower spin dephasing at higher temperatures results from the reduced electron hole exchange interaction - which is a very efficient source for spin relaxation [47,48] - due to thermal ionization of excitons.

We conclude that by intelligent engineering of future semiconductor spin devices spin dephasing can be adjusted at room temperature over several orders of magnitude from extremely short up to very long spin dephasing times. In contrast to spin injection, spin transport at room temperature is therefore in principal a solved problem.

4   Towards Applications

As soon as spin injection at room temperature is feasible the race towards commercial spintronic devices will start. The first device could be a very simple spin-polarized field effect transistor (FET) as shown in Fig. 3. Spin-polarized

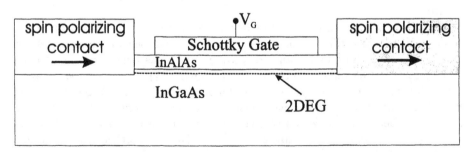

Fig. 3. Spin field-effect transistor according to Datta and Das.

electrons are injected and collected at spin-polarizing contacts, and the current flows, e.g., in an accumulation layer formed at the heterojunction between In-AlAs and InGaAs. Applying an electric field perpendicular to the heterojunction induces an interface spin-orbit effect and rotates the electron spin (Rashba effect). Depending on the magnitude of the electric field the electron spin at the collecting contact is either parallel or antiparallel to the spin polarization of the injecting contact. Therefore, the current is high or low, respectively. More elaborated devices will aim towards the telecommunication market. Not only spin modulated VCSELs are waiting in the starting holes but many spin optoelectronic devices. The matching piece to the spin modulated VCSEL is, for example, a light-polarization sensitive detector which can be build by the same scheme as the VCSEL: circularly polarized light creates spin-polarized electrons which are detected by two electrical spin sensitive contacts. The combination of both could be used for example for faster data transfer or optical data storage. Spin electronics in metals plays a major role in magnetic field sensors, reader heads of computer hard disks, and in the future perhaps even in the DRAM market. To capture or re-capture these fields by semiconductors is another aim of spintronics. The strength of spintronics will be, e.g., the simple integration of detection and complex data processing unit in one tiny device. Corresponding experiments concerning the interlayer exchange in semiconductor ferromagnet/nonmagnet/ferromagnet structures already started.[49,50]

The most futuristic application of spintronics is a solid-state semiconductor spin-quantum-computer. Of course, quantum computers face enormous problems and their implementation is like crasping to the stars but let's summarize the ingredients spintronics offers: (a) Electron and hole spins are quantum mechanical systems which can represent quantum bits. (b) Electrical injection of spin-polarized electrons and holes is feasible. (c) Electron spins and localized hole spins exhibit long coherence times. (d) Electron and hole spin are coupled by the electron hole exchange interaction and can be decoupled by an electric field. (e) The orientation of the spins can be controlled by electrical and magnetic fields. (f) Long-range interaction can be mediated by high finesse microcavities, and (g) parallel controlled-not operations and arbitrary single qubit rotations can in principle be realized. However, while spintronic quantum computers probably will remain a dream, other spintronic devices might crash the gate towards commercial realization in the near future.

5   Acknowledgement

We thank the German Science Foundation (DFG) and the Bundesministerium für Bildung und Forschung (BMBF) for financial support.

Correspondence should be addressed to M.O.
(email: Michael.Oestreich@nano.uni-hannover.de).

References

1. P. Grünberg, R. Schreiber, Y. Pang, M. B. Brodsky, and H. Sowers, Layered Magnetic Structures: Evidence for antiferromagnetic coupling of Fe-layers across Cr-interlayers, Phys. Rev. Lett. **57**, 2442-2444 (1986).
2. R. Schad, et al., Giant magnetoresistance in Fe/Cr superlattices with very thin Fe layers, Appl. Phys. Lett. **64**, 3500-3502 (1994).

3. S. S. P. Parkin, et al., Exchange-biased magnetic tunnel junctions and application to nonvolatile magnetic random access memory, J. Appl. Phys. **85**, 5828-5833 (1999).
4. Gary A. Prinz, Magnetoelectronics. Science **282**, 1660-1663 (1998).
5. M. J. Snelling, et al., Magnetic g factor of electrons in $GaAs/Al_xGa_{1-x}As$ quantum wells, Phys. Rev. B **44**, 11345-11352 (1991); R. M. Hannak, M. Oestreich, A. P. Heberle, and W. W. Rühle Electron g Factor in Quantum Wells Determined by Spin Quantum Beats, Sol. State Comm. **93**, 313-317 (1995); M. Oestreich, et al., Direct Observation of the Rotational Direction of Electron Spin Precession in Semiconductors, Sol. State Comm. **108**, 753-758 (1998).
6. S. Hallstein, et al., Manifestation of Coherent Spin Precession in Stimulated Semiconductor Emission Dynamics, Phys. Rev. B **56**, R7076-R7099 (1997).
7. S. A. Crooker, J. J. Baumberg, F. Flack, N. Samarth, and D. D. Awschalom, Terahertz Spin Precession and Coherent Transfer of Angular Momenta in Magnetic Quantum Wells, Phys. Rev. Lett. **77**, 2814-2817 (1996).
8. G. A. Prinz, Spin-Polarized Transport, Phys. Today **48**, 58-63 (1995).
9. S. F. Alvarado and P. Renaud, Observation of Spin-Polarized-Electron Tunneling from a Ferromagnet into GaAs, Phys. Rev. Lett. **68**, 1387-1379 (1992).
10. P. R. Hammar, B. R. Bennett, M. J. Yang, and M. Johnson, Observation of Spin Injection at a Ferromagnet-Semiconductor Interface, Phys. Rev. Lett. **83**, 203-206 (1999).
11. M. Johnson, Theory of spin-dependent transport in ferromagnet-semiconductor heterostructures, Phys. Rev. B. **58**, 9635-9638 (1998).
12. W. Y. Lee, et al., Magnetization reversal and magnetoresistance in a lateral spin-injection device, J. Appl. Phys. **85**, 6682-6685 (1999).
13. S. Gardelis, et al., Spin-valve effects in a semionductor field-effect transistor: A spintronic device, Phys. Rev. B **60**, 7764-7767 (1999).
14. S. Datta and B. Das, Electronic analog of the electro-optic modulator, Appl. Phys. Lett. **56**, 665-667 (1990).
15. F. G. Monzon, H. X. Tang, and M.L. Roukes, Magnetoelectronic Phenomena at a Ferromagnet-Semiconductor Interface, Phys. Rev. Lett. **84**, 5022 (2000).
16. B. van Wees Comment on "Observation on Spin Injection at a Ferromagnet-Semiconductor Interface" Phys. Rev. Lett. **84**, 5023 (2000).
17. G. Schmidt, D. Ferrand, L. W. Molenkamp, A.T. Filip, B. J. van Wees, Fundamental obstacle for electrical spin-injection from a ferromagnetic metal into a diffusive semiconductor, Phys. Rev. B **62** R4790-R4793 (2000).
18. R. Meservey and P. M. Tedrow, Spin Polarization of Electron Tunneling from Films of Fe, Co, Ni, and Gd, Phys. Rev. B **7**, 318-326 (1973); R. Meservey, D. Paraskevopoulos, P. M. Tedrow, Correlation between spin polarization of tunnel currents from 3d ferromagnets and their magnetic moments, Phys. Rev. Lett. **37**, 858-860 (1976).
19. G. Meier and T. Matsuyama, Magnetic electrodes for spin-polarized injection into InAs, Appl. Phys. Lett. **76**, 1315-1317 (2000).
20. M. Zöfl, et al., Magnetic films epitaxially grown on semiconductors, J. Mag. Mag. Mat. **175**, 16-22 (1997).
21. A. T. Filip, Experimental search for the electrical spin injection in a semiconductor, Phys. Rev. B **62**, 9996-9999 (2000).
22. M. Oestreich, et al., Spin injection into semiconductors, Appl. Phys. Lett. **74**, 1251-1253 (1999).
23. R. Fiederling, et al., Injection and detection of a spin-polarized current in a light-emitting diode, Nature **402**, 787-790 (1999).

24. B. T. Jonker, et al., Robust Electrical Spin Injection into a Semiconductor Heterostructure, Phys. Rev. B **62**, 8180-8183 (2000).
25. J. K. Furdyna, Diluted magnetic semiconductors, J. Appl. Phys. **64**, R29-R64 (1988).
26. Y. Ohno, et al., Electrical Spin Injection in a Ferromagnetic Semiconductor Heterostructure, Nature **402**, 790-792 (1999).
27. H. Ohno, et al., Ferromagnetic Order in (Ga,Mn)As/GaAs Heterostructures, in Proceedings 23rd International Conference on Physics of Semiconductors (World Scientific, Singapur, 1996) 405-408; H. Ohno et al., (Ga,Mn)As: A new diluted magnetic semiconductor based on GaAs, Appl. Phys. Lett. **69**, 363-365 (1996); H. Ohno, F. Matsukura, T. Omiya, and N. Akiba, Spin-dependent tunneling and properties of ferromagnetic (Ga,Mn)As, J. Appl. Phys. **85**, 4277-4282 (1999).
28. T. Dietl, H. Ohno, F. Matsukura, J. Cibert, and D. Ferrand, Zener Model Description of Ferromanetism in Zinc-Blende Magnetic Semiconductors, Science **287**, 1019-1021 (2000).
29. A. R. Cameron, A. R. Riblet, and A. Miller, Determination of the electron mobility in multiple quantum wells by time-resolved optical measurements, Summaries of the Papers Presented at the Topical Meeting Ultrafast Phenomena (Technical Digest Series, ISBN 1 55752 441 6) **8**, 126-127 (1996); S. Adachi, T. Miyashita, S. Takeyama, Y. Takagi, and A. Tackeuchi, Exciton spin dynamics in GaAs quantum wells, J. of Luminescence **72-74**, 307-308 (1997).
30. H. Akinaga, S. Miyanishi, K. Tanaka, W. van Roy, and K. Onodera, Magneto-optical properties and the potential application of GaAs with magnetic MnAs nanoclusters, Appl. Phys. Lett. **76**, 97-99 (2000).
31. H. X. Tang, F. G. Monzon, R. Lifshitz, M. C. Cross, and M. L. Roukes, Ballistic spin transport in a two-dimensional electron gas, Phys. Rev. B **61**, 4437-4440 (2000).
32. J. P. Lu, et al., Tunable Spin-Splitting and Spin-Resolved Ballistic Transport in GaAs/AlGaAs Two-Dimensional Holes, Phys. Rev. Lett. **81**, 1282-1285 (1998).
33. G. Fasol and H. Sakaki, Spontaneous spin polarization of ballistic electrons in single-mode quantum wires due to spin splitting, Appl. Phys. Lett. **62**, 2230-2232 (1993).
34. D. Hägele, M. Oestreich, W. W. Rühle, N. Nestle, and K. Eberl, Spin transport in GaAs, Appl. Phys. Lett. **73**, 1580-1582, (1998).
35. J. M. Kikkawa and D. D. Awschalom, Lateral drag of spin coherence in gallium arsenide, Nature **397**, 139-141 (1998).
36. T. Sogawa, H. Ando, and S. Ando, Spin-transport dynamics of optically spin-polarized electrons in GaAs quantum wires, Phys. Rev. B **61**, 5535-5539 (2000).
37. J. M. Kikkawa and D. D. Awschalom, Resonant Spin Amplification in n-Type GaAs, Phys. Rev. Lett. **80**, 4313-4316 (1998).
38. A. Tackeuchi, O. Wada, and Y. Nishikawa, Electron spin relaxation in InGaAs/InP multiple quantum wells, Appl. Phys. Lett. **70**, 1131-1133 (1997).
39. Y. Nishikawa, A. Tackeuchi, M. Ymaguchi, S. Muto, and O. Wada, Ultrafast All-Optical Spin Polarization Switch Using Quantum-Well Etalon, J. Select. Top. Quantum Electron. **2**, 661-667 (1996).
40. Aronov, A. G., Pikus, G. E., and Titkov, A. N., Spin relaxation of conduction electrons in p-type III-V compounds, Sov. Phys. JETP **57**, 680-687 (1983).
41. S. Adachi, et al., Exciton spin dynamics in GaAs quantum wells, J. Lumin. **72**, 307-308 (1997).
42. M. I. D'yakonov, and V. I. Perel, Spin Orientation of Electrons Associated with the Interband Absorption of Light in Semiconductors, Sov. Phys. JETP **33**, 1053-1059

(1971); Spin Relaxation of Conduction Electrons in Noncentrosymmetric Semiconductors, Sov. Phys. Solid State **13**, 3023-3026 (1972); M. I. D'yakonov and V. Y. Kachorovskii, Spin relaxation of two-dimensional electrons in noncentrosymmetric semiconductors, Sov. Phys. Semicond. **20**, 110-112 (1986).

43. T. Nishimura, X.-L. Wang, M. Ogura, A. Tackeuchi, and O. Wada, Electron spin relaxation in GaAs/AlGaAs quantum wires analysed by transient photoluminescence, Jap. J. Appl. Phys., Part 2:Letters **38**, L941-L944.
44. Y. Ohno, R. Terauchi, T. Adachi, F. Matsukura, and H. Ohno, Spin Relaxation in GaAs(110) Quantum Wells, Phys. Rev. Lett. **83**, 4196-4199 (1999).
45. D. Hägele, et al., Relation between spin and momentum relaxation in ZnSe/ZnMgSSe quantum wells, Physica B **272**, 338-340 (1999).
46. J. M. Kikkawa, I. P. Smorchkova, N. Samarth, and D. D. Awschalom, Room Temperature Spin Memory in Two-Dimensional Electron Gases, Science **277**, 1284-1287 (1997).
47. M. Z. Maialle, E. A. de Andrada e Silva, and L. J. Sham, Exciton spin dynamics in quantum wells, Phys. Rev. B **47**, 15776-15788 (1993).
48. A. Vinattieri, et al., Electric field dependence of exciton spin relaxation in GaAs/AlGaAs quantum wells, Appl. Phys. Lett. **63**, 3164-3166 (1993).
49. N. Akiba, Interlayer exchange in (Ga,Mn)As/(Al,Ga)As/(Ga,Mn)As semiconducting ferromagnet/nonmegnet/ferromagnet trilayer structures, Appl. Phys. Lett. **73**, 2122-2124 (1998).
50. T. Jungwirth, W. A. Atkinson, B. H. Lee, and A. H. MacDonald, Interlayer coupling in ferromagnetic semiconductor superlattices, Phys. Rev. B **59**, 9818-9821 (1999).
51. A. Imamoglu, et al., Quantum Information Processing Using Quantum Dot Spins and Cavity QED, Phys. Rev. Lett. **83**, 4204-4207 (1999).

Theory of Ferromagnetism in Diluted Magnetic Semiconductors

Jürgen König[1,2], Hsiu-Hau Lin[3], and Allan H. MacDonald[1,2]

[1] Department of Physics, Indiana University, Bloomington, IN 47405, USA
[2] Department of Physics, The University of Texas at Austin, Austin, TX 78712, USA
[3] Department of Physics, National Tsing-Hua University, Hsinchu 300, Taiwan

Abstract. Carrier-induced ferromagnetism has been observed in several (III,Mn)V semiconductors. We review the theoretical picture of these ferromagnetic semiconductors that emerges from a model with kinetic-exchange coupling between localized Mn spins and valence-band carriers. We discuss the applicability of this model, the validity of a mean-field approximation for its interaction term widely used in the literature, and validity limits for the simpler RKKY model in which only Mn spins appear explicitly. Our conclusions are based in part on our analysis of the dependence of the system's elementary spin excitations on carrier density and exchange-coupling strength. The analogy between this system and spin-model ferrimagnets is explored. Finally, we list several extensions of this model that can be important in realistic modeling of specific materials.

1   Introduction

Semiconductors and ferromagnets play complementary roles in current information technology. On the one hand, low carrier densities in semiconductors facilitate modulation of electronic transport properties by doping or external gates. Information *processing* is, therefore, most commonly based on semiconductor devices. On the other hand, because of long-range order in ferromagnets only small magnetic fields are necessary to reorient large magnetic moments and induce large changes in magnetic and transport properties. This explains the use of ferromagnetic metals in *storage* devices. The application of giant magnetoresistance (GMR) and tunneling magnetoresistance (TMR) effects to read magnetically stored information is the poster child of spin electronics in ferromagnetic *metals*. The prospect of synergism [1,2] between electronic and magnetic manipulation of transport properties has raised interest in spin electronics in *semiconductors*. In a (III,Mn)V ferromagnetic semiconductor charge and spin degrees of freedom could, it is hoped, be manipulated on the same footing in basically the same material used now in highly successful electronic devices.

Two preconditions for spin electronics in semiconductors are, first, the generation of spin-polarized carriers and, second, their spin-coherent transport in the material. While spin-coherent transport over several micrometers has been demonstrated [3–5], spin injection from ferromagnetic metals into semiconductors has proven to be problematic [6]. More promising is the use of ferromagnetic semiconductors, in which not only charges but also magnetic moments are introduced by doping [7,8].

1.1   Diluted Magnetic II-VI and III-V Semiconductors

Semiconductors have rich optical and magnetic properties when Mn or other magnetic ions are doped into the material [9,10]. For a long time, the investigation of these so-called diluted magnetic semiconductors (DMS) concentrated on II-VI compounds such as CdTe or ZnSe, where the valence of the cation is identical to that of Mn. The magnetic interaction among the Mn spins is in this case, however, dominated by antiferromagnetic direct exchange, and the systems show paramagnetic, antiferromagnetic or spin-glass behavior. To obtain ferromagnetism, free carriers must be introduced as we explain below. This can be achieved by additional doping [11], which is, however, rather difficult experimentally. Only low transition temperatures have been realized by this route so far.

An alternative approach is to dope Mn in III-V semiconductors such as GaAs [12–22]. Ferromagnetic samples with transition temperatures T_c up to 110 K [14,15] have been realized using low-temperature molecular beam epitaxy (MBE) growth, and it is hoped that still higher transition temperatures will be achieved in the future. Due to the difference between the valence of Mn^{2+} and the cation (e.g. Ga^{3+}), free carriers (holes) are then generated in the valence band. The main focus of this article is on the mechanism by which these carriers mediate interactions that lead to long-range ferromagnetic order among the Mn spins.

1.2   Model for Mn-Doped III–V Semiconductors

We use a partially phenomenological approach, starting from a relatively simple but soundly based model with effective parameters that can be determined by comparison with experiment. We explore the physical properties implied by this model. The reliability of the model will ultimately be determined by comparison with future experimental findings. We will start by discussing the simplest version of this model; a variety of extensions can be accommodated without any fundamental changes. Some of the anticipated extensions will be discussed at the end of this article.

It is believed that Mn^{2+} impurities substitute randomly for the Ga^{3+} cation in the zincblende structure. We assume that the Mn ion, with its half-filled d shell, acts as a $S = 5/2$ local moment. If interactions did not suppress fluctuations in the Mn ion charge sufficiently, its d-electrons would have to be treated as itinerant, and this model would fail. We denote the Mn ion positions by R_I. Hybridization between the localized Mn d-orbitals and the valence-band orbitals can then be represented by an antiferromagnetic exchange interaction. It turns out that the net valence-band carrier concentration p is much smaller than the Mn ion density N_{Mn} [15]. This might be due to antisite defects (As ions sitting on a Ga lattice site, providing two free electrons per defect) which are common in low-temperature MBE grown materials. As will become clear, we believe that this property of the as-grown materials is actually essential to the robustness of their ferromagnetism. These considerations suggest the following model

Hamiltonian:

$$H = H_0 + J_{\mathrm{pd}} \int d^3r \; \boldsymbol{S}(\boldsymbol{r}) \cdot \boldsymbol{s}(\boldsymbol{r}), \tag{1}$$

where $\boldsymbol{S}(\boldsymbol{r}) = \sum_I \boldsymbol{S}_I \delta(\boldsymbol{r} - \boldsymbol{R}_I)$ is the impurity-spin density. In the following, we approximate the sum of delta peaks by a smooth function, thus neglecting disorder induced by randomness of Mn sites. This approximation is partly justified by the carrier to impurity density ratio p/N_{Mn}, since the typical distance between Mn ions is smaller than the typical distance between free carriers. The itinerant-carrier spin density is expressed in terms of carrier field operators by $\boldsymbol{s}(\boldsymbol{r}) = \frac{1}{2} \sum_{\sigma\sigma'} \Psi_\sigma^\dagger(\boldsymbol{r}) \boldsymbol{\tau}_{\sigma\sigma'} \Psi_{\sigma'}(\boldsymbol{r})$ where $\boldsymbol{\tau}$ is the vector of Pauli spin matrices. The antiferromagnetic exchange interaction between valence-band holes and local moments is represented by the exchange integral $J_{\mathrm{pd}} > 0$, an effective parameter which is dependent on the details of the system's atomic length scale physics. The contribution H_0 includes the valence-band envelope-function Hamiltonian [23] and, if an external magnetic field $\boldsymbol{B}$ is present, the Zeeman energy,

$$H_0 = \int d^3r \left\{ \sum_\sigma \hat{\Psi}_\sigma^\dagger(\boldsymbol{r}) \left(-\frac{\hbar^2 \boldsymbol{\nabla}^2}{2m^*} - \mu \right) \hat{\Psi}_\sigma(\boldsymbol{r}) - \mu_B \boldsymbol{B} \cdot [g\boldsymbol{S}(\boldsymbol{r}) + g^*\boldsymbol{s}(\boldsymbol{r})] \right\}. \tag{2}$$

The effective mass, chemical potential, and g-factor of the itinerant carriers are labeled by m^*, μ, and g^*, respectively. To simplify the present discussion we use a generic single-band model with quadratic dispersion.

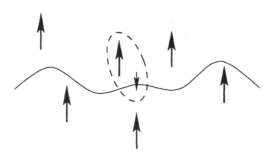

Fig. 1. Model for Mn doped III-V semiconductors: local magnetic moments (Mn^{2+}) with spin $S = 5/2$ are antiferromagnetically coupled to itinerant carriers (holes) with spin $s = 1/2$.

We note that the model we use here is identical to those for dense Kondo systems, which simplify when the itinerant-carrier density p is much smaller than the magnetic ion density N_{Mn} [25]. The fact that $p/N_{\mathrm{Mn}} \ll 1$ (while the Kondo effect shows up in the opposite regime) in ferromagnetic-semiconductor materials is essential to their ferromagnetism. Similar models have been used for $s - f$ materials [26] with ferromagnetic exchange and for ferromagnetism induced by magnetic ions in nearly ferromagnetic metals such as palladium [27].

Additional realism can be included in the model by using a realistic band model, and by accounting for carrier-carrier and carrier-dopant Coulomb interactions.

1.3   Outline

The outline of this article is as follows. In Section 2 we point out that neither the RKKY picture nor the mean-field approximation to the above kinetic-exchange model yield completely satisfactory theories of ferromagnetism in doped DMSs. To develop a more complete picture of these ferromagnets we present in Section 3 a recently developed theory [28–30], in which we identify the elementary spin excitations (Section 4) of the kinetic-exchange model. We compare the excitation spectra that result with those predicted by the RKKY picture (Section 5), by the mean-field approach (Section 6), and by a ferrimagnetic lattice model (Section 7). Section 8 contains an alternative derivation of the spin-wave dispersion, giving also some insight in the nature of the excited states. Implications of these results for the excitation spectrum for other properties of these materials and extensions of the simple model presented above are discussed in Section 9.

2   Origin of Ferromagnetism

2.1   RKKY Interaction

Simple perturbation theory can be used to describe the familiar itinerant-carrier-mediated Ruderman-Kittel-Kasuya-Yoshida (RKKY) interaction between local magnetic moments. At first glance, this interaction might explain ferromagnetism in DMSs, too: each local impurity polarizes the itinerant carriers nearby, an adjacent Mn ion experiences this polarization as an effective magnetic field and aligns, for small distance between the Mn ions, parallel to the first local moment.

The RKKY theory is, however, a (second-order) perturbation expansion in the exchange coupling J_{pd}, i.e., the picture is applicable as long as the perturbation induced by the Mn spins on the itinerant carriers is small. As we will derive below, the proper condition (at zero external magnetic field B) is $\Delta \ll \epsilon_F$ where $\Delta = N_{Mn} J_{pd} S$ is the spin-splitting gap of the itinerant carriers due to an average effective magnetic field induced by the Mn ions, and ϵ_F is the Fermi energy. This condition is, however, never satisfied in (III,Mn)V ferromagnets, partially because (as mentioned in the introduction) the valence-band carrier concentration p is usually much smaller than the Mn impurity density N_{Mn}. Instead, the "perturbation" is strong. Valence-band spin splitting comparable in size to the Fermi energy has been confirmed by direct experiment [31]. *The RKKY description does not provide a good starting point to describe the ordered state in ferromagnetic DMSs.*

A related drawback of the RKKY picture is that it assumes an instantaneous static interaction between the magnetic ions, i.e., the dynamics of the free carriers are neglected. As we will see below this dynamics is important to obtain all types of elementary spin excitations.

Within mean-field theory, however, the spin-splitting of the valence bands will vanish as the critical temperature is approached. It turns out that as a result the RKKY picture and the mean-field-theory approximation to the kinetic-exchange model make identical predictions [32] for the critical temperature. We comment later on the accuracy of these critical temperature estimates.

2.2 Mean-Field Picture

The tendency toward ferromagnetism and trends in the observed T_c's have been explained using a mean-field approximation [32–37], analogous to the simple Weiss mean-field theory for lattice spin models. Each local magnetic moment is treated as a free spin in an effective external field generated by the mean polarization of the free carriers. Similarly, the itinerant carriers see an effective field proportional to the local moment density and polarization. At zero temperature, the total energy is minimized by a state in which all impurity spins are oriented in the same direction, which we define as the positive z-axis, and the itinerant-carrier spins are aligned in the opposite direction. The spin-splitting gap (at zero external magnetic field B) for the free carriers is $\Delta = N_{Mn}J_{pd}S$, and the energy gap for an impurity-spin excitation is $\Omega^{MF} = p\xi J_{pd}/2$, where $\xi = (p_\downarrow - p_\uparrow)/p$ is the fractional free-carrier spin polarization.

This picture, however, neglects correlation between local-moment spin configurations and the free-carrier state and, therefore, fails to describe the existence of low-energy long-wavelength spin excitations. As we discuss below, correlation always lowers the energy Ω_k of collective spin excitations (spin waves) in comparison to the mean-field value Ω^{MF}. Goldstone's theorem even requires the existence of a soft mode for the isotropic model we study here; its absence is the most serious failure of mean-field theory. Because of its neglect of collective magnetization fluctuations, mean-field theory also always overestimates the critical temperature [28,38].

2.3 New Theory

In this article we present a theory [28–30] of DMS ferromagnetism that accounts for both finite itinerant-carrier spin splitting and dynamical correlations. It goes, thus, beyond the RKKY picture and mean-field theory and allows us to analyze the ordered state and its fundamental properties. The starting point for this new theory is an analysis of the system's collective excitations that we discuss in the following sections.

3   Derivation of Independent Spin-Wave Theory

In this section we use a path-integral formulation and derive, after integrating out the itinerant carriers, an effective action for the Mn spin system. We expand the action up to quadratic order to obtain the propagator for independent spin excitations. The latter will be the starting point for Section 4 where we identify the system's elementary spin excitations and their dispersion.

3.1   Effective Action

We represent the impurity-spin density in terms of Holstein-Primakoff (HP) bosons [39],

$$S^+(r) = \left(\sqrt{2N_{\mathrm{Mn}}S - b^\dagger(r)b(r)} \right) b(r) \qquad (3)$$

$$S^-(r) = b^\dagger(r)\sqrt{2N_{\mathrm{Mn}}S - b^\dagger(r)b(r)} \qquad (4)$$

$$S^z(r) = N_{\mathrm{Mn}}S - b^\dagger(r)b(r) \qquad (5)$$

with bosonic fields $b^\dagger(r), b(r)$. As can be seen from Eq. (5), the state with fully polarized Mn spins (along the z-direction, which we chose as the quantization axis) corresponds, in the HP boson language, to the vacuum with no bosons. The creation of an HP boson reduces the magnetic quantum number by one. The square roots in Eqs. (3) and (4) restrict the boson Hilbert space to the physical subspace with at most $2S$ bosons per Mn spin. This constraint obviously yields an interaction between spin excitations. Later, when we go to the independent spin-wave theory, we will approximate the square roots by $\sqrt{2N_{\mathrm{Mn}}S}$, i.e., we will treat the spin excitations as free bosons. This is a good approximation as long as the spin-excitation density is small.

We can write down the partition function of our model as a coherent-state path-integral in imaginary times,

$$Z = \int \mathcal{D}[\bar{z}z]\mathcal{D}[\bar{\Psi}\Psi]e^{-\int_0^\beta d\tau L(\bar{z}z, \bar{\Psi}\Psi)} \qquad (6)$$

with the Lagrangian $L = \int d^3r \left[\bar{z}\partial_\tau z + \sum_\sigma \bar{\Psi}_\sigma \partial_\tau \Psi_\sigma \right] + H(\bar{z}z, \bar{\Psi}\Psi)$. The bosonic (impurity spin) and fermionic (itinerant carrier) degrees of freedom are represented by the complex and Grassmann number fields, $\bar{z}, z$ and $\bar{\Psi}, \Psi$, respectively.

Since the Hamiltonian is bilinear in fermionic fields, we can integrate out the itinerant carriers and arrive at an effective description in terms of the localized spin density only, $Z = \int \mathcal{D}[\bar{z}z] \exp(-S_{\mathrm{eff}}[\bar{z}z])$ with the effective action

$$S_{\mathrm{eff}}[\bar{z}z] = \int_0^\beta d\tau \int d^3r \left[\bar{z}\partial_\tau z - g\mu_B B(N_{\mathrm{Mn}}S - \bar{z}z) \right]$$
$$- \ln \det \left[(G^{\mathrm{MF}})^{-1} + \delta G^{-1}(\bar{z}z) \right] . \qquad (7)$$

Here, we have already split the total kernel G^{-1} into a mean-field part $(G^{\mathrm{MF}})^{-1}$, which does not depend on the fields z and $\bar{z}$, and a fluctuating part δG^{-1},

$$(G^{\mathrm{MF}})^{-1} = \left(\partial_\tau - \frac{\hbar^2 \nabla^2}{2m^*} - \mu \right) 1 + \frac{\Delta}{2}\tau^z , \qquad (8)$$

$$\delta G^{-1} = \frac{J_{\mathrm{pd}}}{2} \left[(z\tau^- + \bar{z}\tau^+)\sqrt{2N_{\mathrm{Mn}}S - \bar{z}z} - \bar{z}z\tau^z \right] , \qquad (9)$$

where $\Delta = N_{\mathrm{Mn}}J_{\mathrm{pd}}S - g^*\mu_B B$ is the zero-temperature spin-splitting gap for the itinerant carriers. We have allowed for the possibility of an external magnetic field B along the ordered moment direction. Although the itinerant carriers have been integrated out, their physics is still embedded in the effective action of the magnetic ions. This fact is responsible for the retarded and non-local character of the interactions between magnetic ions.

3.2 Independent Spin-Wave Theory

The independent spin-wave theory is obtained by expanding Eq. (7) up to quadratic order in z and $\bar{z}$, i.e., spin excitations are treated as noninteracting HP bosons. This is a good approximation at low temperatures, where the number of spin excitations per Mn site is small.

We expand the term $\ln \det \left[(G^{\mathrm{MF}})^{-1} + \delta G^{-1} \right]$ up to second order in δG^{-1}, $\ln \det \left(G^{-1} \right) = \operatorname{tr} \ln \left(G^{\mathrm{MF}} \right)^{-1} + \operatorname{tr} \left(G^{\mathrm{MF}} \delta G^{-1} \right) - \frac{1}{2} \operatorname{tr} \left(G^{\mathrm{MF}} \delta G^{-1} G^{\mathrm{MF}} \delta G^{-1} \right)$, and collect all contributions up to quadratic order in z and $\bar{z}$. The diagrammatic representation of the contribution linear and quadratic in δG^{-1} are depicted in Fig. 2. We obtain (in the imaginary time Matsubara and coordinate Fourier

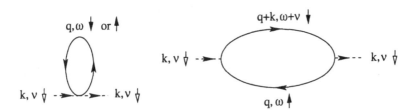

Fig. 2. Diagrammatic representation of $\operatorname{tr} \left(G^{\mathrm{MF}} \delta G^{-1} \right)$ and $\operatorname{tr} \left(G^{\mathrm{MF}} \delta G^{-1} G^{\mathrm{MF}} \delta G^{-1} \right)$. The solid lines represent the mean-field Green's function for the itinerant carriers and the incoming and outgoing dashed lines stand for z and $\bar{z}$, respectively.

representation) an action that is the sum of the temperature-dependent mean-field contribution and a fluctuation action,

$$S_{\mathrm{eff}}[\bar{z}z] = \frac{1}{\beta V} \sum_{|\mathbf{k}|<k_D,m} \bar{z}(\mathbf{k},\nu_m) D^{-1}(\mathbf{k},\nu_m) z(\mathbf{k},\nu_m). \qquad (10)$$

A Debye cutoff k_D with $k_D^3 = 6\pi^2 N_{\mathrm{Mn}}$ ensures that we include the correct number of magnetic-ion degrees of freedom, $|\mathbf{k}| \leq k_D$. The kernel of the quadratic action defines the inverse of the spin-wave propagator,

$$D^{-1}(\mathbf{k},\nu_m) = -i\nu_m + g\mu_B B + \frac{J_{\mathrm{pd}} p \xi}{2}$$
$$+ \frac{N_{\mathrm{Mn}} J_{\mathrm{pd}}^2 S}{2\beta V} \sum_{n,q} G_\uparrow^{\mathrm{MF}}(\mathbf{q},\omega_n) G_\downarrow^{\mathrm{MF}}(\mathbf{q}+\mathbf{k},\omega_n+\nu_m), \qquad (11)$$

where $G_\sigma^{MF}(q, \omega_n)$ is the mean-field itinerant carrier Green's function given by $G_\sigma^{MF}(q, \omega_n) = -[i\omega_n - (\epsilon_q + \sigma\Delta/2 - \mu)]^{-1}$, and $\epsilon_q = \hbar^2 q^2/(2m^*)$. The fractional free-carrier spin polarization is $\xi = (p_\downarrow - p_\uparrow)/p$.

4   Elementary Spin Excitations

We obtain the spectral density of the spin-fluctuation propagator by analytical continuation, $i\nu_m \to \Omega + i0^+$ and $A(p, \Omega) = \mathrm{Im}\, D(p, \Omega)/\pi$. In the following we consider the case of zero external magnetic field, $B = 0$, and zero temperature, $T = 0$. We find three different types of spin excitations.

4.1   Goldstone-Mode Spin Waves

Our model has a gapless Goldstone-mode branch reflecting the spontaneous breaking of spin-rotational symmetry. At long-wavelengths, the free-carrier and magnetic-ion spin-density orientation variation is identical in these modes. The dispersion of this low-energy mode for four different valence-band carrier concentrations p is shown in Fig. 3 (solid lines). At large momenta, $k \to \infty$, the spin-wave energy approaches the mean-field result (short-dashed lines in Fig. 3)

$$\Omega_{k\to\infty}^{(1)} = \Omega^{MF} = x\Delta. \tag{12}$$

Note that the itinerant-carrier and magnetic-ion mean-field spin splittings differ by the ratio of the spin densities

$$x = p\xi/(2N_{Mn}S); \tag{13}$$

x is always much smaller than 1 in (III,Mn)V ferromagnets. Expansion of the $T = 0$ propagator for small momenta yields for the collective modes dispersion,

$$\Omega_k^{(1)} = \frac{x/\xi}{1-x}\epsilon_k \left(\frac{3+2\xi}{5} - \frac{4}{5}\xi\frac{\epsilon_F}{\Delta}\right) + \mathcal{O}(k^4), \tag{14}$$

where ϵ_F is the Fermi energy of the majority-spin band. The spectral weight of these modes at zero momentum is $1/(1-x)$. In strong and weak-coupling limits, $\Delta \gg \epsilon_F$ and $\Delta \ll \epsilon_F$, respectively, Eq. (14) simplifies to

$$\Omega_k^{(1)} = \frac{x}{1-x}\epsilon_k + \mathcal{O}(k^4) \qquad \text{for} \qquad \Delta \gg \epsilon_F, \tag{15}$$

$$\Omega_k^{(1)} = \frac{p}{32N_{Mn}S}\epsilon_k \left(\frac{\Delta}{\epsilon_F}\right)^2 + \mathcal{O}(k^4) \qquad \text{for} \qquad \Delta \ll \epsilon_F. \tag{16}$$

At long wavelengths, the magnon dispersion $\Omega_k = \rho k^2/M$ in an isotropic ferromagnet is proportional the spin stiffness ρ divided by the magnetization M, in our case $M = N_{Mn}S - p\xi/2$. Note that the spin stiffness depends differently on the exchange constant J_{pd}, effective mass m^*, and the concentrations N_{Mn}

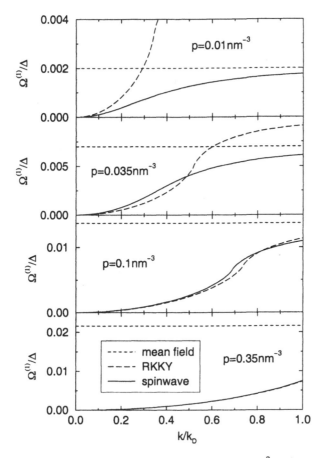

Fig. 3. Spin-wave dispersion (solid lines) for $J_{pd} = 0.06\text{eVnm}^3$, $m^* = 0.5m_e$, $N_{Mn} = 1\text{nm}^{-3}$, and four different itinerant-carrier concentrations $p = 0.01\,\text{nm}^{-3}$, $0.035\,\text{nm}^{-3}$, $0.1\,\text{nm}^{-3}$, and $0.35\,\text{nm}^{-3}$. The ratio Δ/ϵ_F is 2.79, 1.21, 0.67, and 0.35, which yields the fractional free-carrier spin polarization ξ as 1, 1, 0.69, and 0.31. The short wavelength limit is the mean-field result $\Omega^{MF} = x\Delta$ (short-dashed lines). For comparison, we show also the result obtained from an RKKY picture (long-dashed lines).

and p in the strong and weak-coupling limits. The smaller the spin stiffness, the smaller the cost in energy to vary the spin orientation across the sample. A small spin stiffness will limit the temperature at which long-range order in the magnetization orientation, i.e. ferromagnetism, can be maintained. We see from Eq. (15) that the spin stiffness becomes small in systems with large effective band masses (such that $\Delta \gg \epsilon_F$), the same circumstances under which the mean-field-theory critical temperature becomes large. It follows that a mean-field description becomes more and more inappropriate [28,38] in systems with larger itinerant-carrier densities of states. This observation is important in devising strategies for finding (III,Mn)V ferromagnets with larger critical temperatures. We will discuss this point in more detail in Section 8.

4.2   Stoner Continuum

We find a continuum of Stoner spin-flip particle-hole excitations. They correspond to flipping a single spin in the itinerant-carrier system and, since $x \ll 1$, occur at much larger energies near the itinerant-carrier spin-splitting gap Δ (see Fig. 4). For $\Delta > \epsilon_F$ and zero temperature, all these excitations carry spin $S^z = +1$, i.e., increase the spin polarization. They therefore turn up at negative frequencies in the boson propagator we study. (When $\Delta < \epsilon_F$, excitations with both $S^z = +1$ and $S^z = -1$ contribute to the spectral function.) This continuum lies between the curves $-\Delta - \epsilon_k \pm 2\sqrt{\epsilon_k \epsilon_F}$ and for $\Delta < \epsilon_F$ also between $-\Delta + \epsilon_k \pm 2\sqrt{\epsilon_k(\epsilon_F - \Delta)}$.

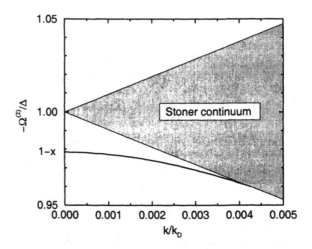

Fig. 4. Stoner excitations and optical spin-wave mode in the free-carrier system for $J_{\mathrm{pd}} = 0.06\mathrm{eVnm}^3$, $m^* = 0.5 m_e$, $N_{\mathrm{Mn}} = 1\mathrm{nm}^{-3}$, and $p = 0.35\mathrm{nm}^{-3}$. In an RKKY picture these modes are absent.

4.3   Optical Spin Waves

We find additional collective modes analogous to the optical spin waves in a ferrimagnet. Their dispersion lies below the Stoner continuum (see Fig. 4). At small momenta the dispersion is

$$-\Omega_k^{(2)} = \Delta(1 - x) - \frac{\epsilon_k}{1 - x}\left(\frac{4\epsilon_F}{5x\Delta} - \frac{2 - (2 - 5x)/\xi}{5x}\right) + \mathcal{O}(k^4). \qquad (17)$$

The spectral weight of these modes is $-x/(1-x)$ at zero momentum.

The finite spectral weight at negative energies indicates that, because of quantum fluctuations, the ground state is not fully spin polarized.

5   Comparison to RKKY Picture

For comparison we evaluate the $T = 0$ magnon dispersion assuming an RKKY interaction between magnetic ions. This approximation results from our theory if we neglect the spin polarization in the itinerant carriers and evaluate the static limit of the resulting spin-wave propagator defined in Eq. (11). The Stoner excitations and optical spin waves shown in Fig. 4 are then not present and the Goldstone-mode dispersion is incorrect except when $\Delta \ll \epsilon_F$, as depicted in Fig. 3 (long-dashed lines). As a conclusion, the RKKY picture can be applied in the weak-coupling regime only (in this limit the long-wavelength fluctuations are described by Eq. (16)). Outside that regime it completely fails as a theory of the ferromagnetic state.

6   Comparison to Mean-Field Picture

In the mean-field picture, correlations among the Mn spins are neglected. The mean-field theory can be obtained in our approach by taking the Ising limit, i.e., replacing $\boldsymbol{S} \cdot \boldsymbol{s}$ by $S^z s^z$. This amounts to dropping the last term in Eq. (11) or the second diagram in Fig. 2. It is this term that describes the response of the free-carrier system to changes in the magnetic-ion configuration. When it is neglected, the energy of an impurity-spin excitation is dispersionless, $\Omega^{\mathrm{MF}} = x\Delta$, and always larger than the real spin-wave energy as can be seen in Fig. 3 (the short dashed line is the mean-field value and the solid line shows the real spin-wave energy).

However, if both the weak-coupling ($\Delta/\epsilon_F \ll 1$) and dilute-density ($p/N_{\mathrm{Mn}} \ll 1$) limit are met, the spin-wave energy is almost dispersionless $\Omega_{\boldsymbol{k}} \approx \Omega^{\mathrm{MF}}$ (see Fig. 5), except in the narrow window near $\boldsymbol{k} = 0$ which is protected by Goldstone's theorem. In this regime, the spatial correlations among Mn spins are less important.

7   Comparison to a Ferrimagnet

Some features of the excitation spectrum are, not coincidentally, like those of a localized spin ferrimagnet with antiferromagnetically coupled large spin S (corresponding to the magnetic ions) and small spin s (corresponding to the itinerant carriers) subsystems on a bipartite cubic lattice (see Fig. 6). We can represent the spins by HP bosons and expand up to quadratic order. Then we either diagonalize the resulting Hamiltonian directly using a Bogoliubov transformation or, as in our ferromagnetic-semiconductor calculations, integrate out the smaller spins using a path-integral formulation. The latter approach is less natural and

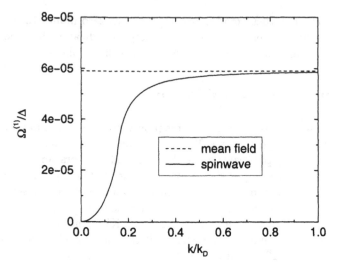

Fig. 5. Spin-wave dispersion (solid line) for $\Delta/\epsilon_F = 1/3$ and $p/N_{\mathrm{Mn}} = 0.001$, in comparison to the mean-field value Ω^{MF} (short-dashed line).

unnecessary in the simple lattice model case, but leads to equivalent results. In both cases we find that there are two collective modes with dispersion

$$\Omega_{\bm{k}}^{(1)/(2)} = \frac{\Delta}{2}\left[-(1-x) \pm \sqrt{(1-x)^2 + 4x\gamma_{\bm{k}}}\right] . \tag{18}$$

In analogy to the ferromagnetic DMS, we defined the ratio of the spins as $x = s/S$, the mean-field spin-splitting gap is $\Delta = 6JS$ for the smaller spins, where J is the exchange coupling, and $\gamma_{\bm{k}} = (1/3)\sum_i[1 - \cos(k_i a)]$ with lattice constant a. The two collective modes correspond to coupled spin waves of the two subsystems. One is gapless, the other one gapped with $\Delta(1 - x)$, exactly as for ferromagnetic semiconductors. The bandwidth is $x\Delta$, and the spectral weights at zero momentum are $1/(1 - x)$ and $-x/(1 - x)$, respectively, as in our model.

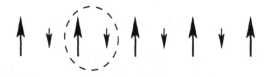

Fig. 6. Localized spin ferrimagnet. Local moments with large spin S and small spin s are sitting on a bipartite cubic lattice. They are coupled by a nearest-neighbor antiferromagnetic interaction. Only one dimension is shown.

This comparison demonstrates that some of our results for the collective excitations of ferromagnetic semiconductors do not depend on details of the model but reflect generic features of a system in which two different types of spins are antiferromagnetically coupled.

8   Alternative Derivation of the Spin-Wave Dispersion

The energy of long-wavelength spin waves is determined by a competition between exchange and kinetic energies. To understand this in detail we impose the spin configuration of a static spin wave with wavevector k on the Mn spin system (see Fig. 7), evaluate the ground-state energy of the itinerant-carrier system in the presence of this exchange field, and compare this with the ground-state energy of a uniformly polarized system. We will rederive the dispersion Ω_k given in Eq. (14) as the ratio of energy increase δE and decrease of the total spin δS.

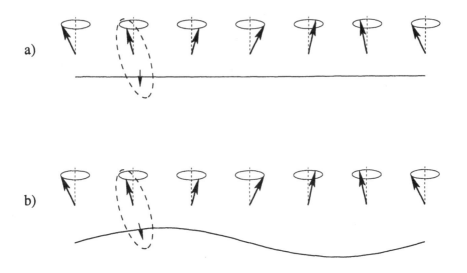

Fig. 7. a) To minimize the kinetic energy the itinerant carriers align antiparallel to the average Mn spin orientation. In this case, exchange energy is increased. b) To minimize the exchange energy the itinerant carriers align antiparallel to the Mn spin orientation everywhere in space. In this case, kinetic energy is increased.

How will the itinerant carriers respond to the inhomogeneous Mn spin orientation? On the one hand, they can minimize their kinetic energy by forming a homogeneously polarized state with the spin pointing in negative z-direction (see Fig. 7a). In this case, however, the Mn and the itinerant-carrier spins are not aligned antiparallel, i.e., the exchange energy is not optimized. One the other hand, if the valence-band spins decide to follow the Mn spin orientation everywhere in space, as indicated in Fig. 7b, then the exchange energy is minimized but the kinetic energy is increased because of a new contribution to spatial variation in the itinerant-carrier spinors. Since the energy scales of the exchange and kinetic energy are provided by Δ and ϵ_F, respectively, it is the ratio Δ/ϵ_F that determines which term controls the itinerant-carrier behavior.

8.1   Strong-Coupling Limit

In the strong-coupling regime, $\Delta/\epsilon_F \gg 1$, the itinerant-carrier spins minimize the exchange energy by following the Mn spin orientation in space as shown in Fig. 7b. The energy δE will therefore be purely kinetic and the spin-wave energy will not depend on Δ in this regime. To evaluate the kinetic energy cost we use a variational ansatz with trial single-particle spinors

$$|\Psi(\boldsymbol{q})\rangle_a = e^{iax} \begin{pmatrix} \sin(\theta/2)e^{+ikx/2} \\ -\cos(\theta/2)e^{-ikx/2} \end{pmatrix} e^{i\boldsymbol{q}\cdot\boldsymbol{r}}, \tag{19}$$

for a free carrier with wavevector $\boldsymbol{q}$, where k is the wavevector of the spin wave going in x-direction, and θ is the angle by which the Mn spins are tilted from their mean-field orientation. The total kinetic energy, i.e., the sum of the single-particle kinetic energies $E_{\mathrm{kin},a}(\boldsymbol{q}) = \langle \Psi(\boldsymbol{q})| - \hbar^2\nabla^2/(2m^*)|\Psi(\boldsymbol{q})\rangle_a$, is minimized for $a_{\min} = (k/2)\cos\theta$, which yields for the increase in kinetic energy δE due to the spin wave

$$\frac{\delta E}{V} = \frac{1}{4}p\epsilon_k \sin^2\theta. \tag{20}$$

We note that

$$\frac{\delta S}{V} = (N_{\mathrm{Mn}}S - p/2)(1 - \cos\theta) \tag{21}$$

and obtain

$$\Omega_k^{(1)} = \frac{x}{1-x}\epsilon_k \cos^2(\theta/2). \tag{22}$$

In the limit of small spin-excitation density, $\theta \to 0$, we recover the independent spin-wave theory result, Eq. (15).

8.2   General Case

We now generalize this discussion to arbitrary values of Δ/ϵ_F. The Hamiltonian for itinerant carriers in an effective potential generated by the Mn spins is given in second quantization by

$$H = \sum_q \left(c^\dagger_{q-\frac{k}{2},\uparrow} \; c^\dagger_{q+\frac{k}{2},\downarrow} \right) \begin{pmatrix} \epsilon_{q-\frac{k}{2}} + (\Delta/2)\cos\theta & (\Delta/2)\sin\theta \\ (\Delta/2)\sin\theta & \epsilon_{q+\frac{k}{2}} - (\Delta/2)\cos\theta \end{pmatrix} \begin{pmatrix} c_{q-\frac{k}{2},\uparrow} \\ c_{q+\frac{k}{2},\downarrow} \end{pmatrix}$$

$$= \sum_{q,\pm} E_{q,\pm} a^\dagger_{q,\pm} a_{q,\pm} \tag{23}$$

with quasiparticle energies

$$E_{q,\pm} = \epsilon_{q\mp\frac{k}{2}}\cos\theta + \epsilon_{\frac{k}{2}}\sin\theta \pm \left[\frac{\Delta}{2} + \frac{\sin^2\theta}{\Delta} \left(\frac{\hbar^2(\boldsymbol{k}\cdot\boldsymbol{q})}{2m} \right)^2 \right] + \mathcal{O}(k^4). \tag{24}$$

It is now straightforward to calculate the band energy of the new quasiparticle bands. In the strong-coupling limit, only the lower quasiparticle band is occupied, the only term from Eq. (24) which enters the energy difference δE is $\epsilon_{\frac{k}{2}\sin\theta}$, and we recover Eq. (20). For arbitrary values of Δ/ϵ_F we arrive after a lengthy calculation at

$$\frac{\delta E}{V} = \frac{1}{4} p\epsilon_k \sin^2\theta \left(\frac{3+2\xi}{5} - \frac{4}{5}\xi\frac{\epsilon_F}{\Delta} \right) + \mathcal{O}(k^4) \tag{25}$$

which yields together with

$$\frac{\delta S}{V} = (N_{\mathrm{Mn}}S - p\xi/2)(1 - \cos\theta) \tag{26}$$

the desired result (compare to Eq. (14))

$$\Omega_k = \frac{x/\xi}{1-x}\epsilon_k \left(\frac{3+2\xi}{5} - \frac{4}{5}\xi\frac{\epsilon_F}{\Delta} \right) \cos^2(\theta/2) + \mathcal{O}(k^4). \tag{27}$$

9   Extensions of the Model

The discussion in preceding sections has been based on a number of assumptions and idealizations that have enabled us to expose some essential aspects of the physics in a relatively simple way. Among these we mention first the assumption that Mn ion charge fluctuations are weak and therefore can be described perturbatively, representing their influence by effective exchange interactions with itinerant carriers. Failure of this assumption would completely invalidate the approach taken here. In our judgement, however, experimental evidence is completely unambiguous on this issue and favors the approach we have taken. Another essential assumption we have made is that we can use an effective-mass approximation, or more generally an envelope-function approach, to describe the influence of the kinetic-exchange interaction on the itinerant bands. This assumption is safe provided that itinerant-carrier wavefunctions are not distorted on atomic length scales by their interactions with the Mn acceptors. We believe that effective-mass-theory's success [24] in interpreting isolated Mn acceptor binding energy data, albeit with central cell corrections (see below) not included in the simplest model, establishes its applicability in (III,Mn)V semiconductors. Many other extensions of our simplest model that may be required for experimental realism can be included without any essential change in the structure of the theory. We mention some of these below.

- The valence bands are p-type, with 6 orbitals including the spin-degree of freedom. In addition, they are not isotropic, but reflect the cubic symmetry of the crystal. A more realistic description is provided by a six-band Kohn-Luttinger Hamiltonian [40] that includes an essential spin-orbit coupling term and is parametrized by phenomenological constants whose values have been accurately determined by experiment. Furthermore, depending on the substrate on which the sample is grown, either compressive or tensile strain modifies the band structure. This effect can also be included in the Kohn-Luttinger Hamiltonian.
  Mean-field calculations based on this more realistic band structure have been performed recently [34,36]. Since the model is no longer isotropic the spin-wave dispersion will be gapped.
- The impurity density is not continuous. Instead, discrete Mn ions are randomly distributed on cation lattice sites. In recent numerical finite-size calculations we have found [38] that this disorder has important effects. It leads, e.g., to an increase of the spin stiffness in the strong-coupling regime and presumably also plays an important role in determining the $T = 0$ resistivity.
- Evidence from studies of (II,Mn)VI semiconductors suggests that direct antiferromagnetic exchange between Mn ions is important when they are on neighboring sites. The way in which this interaction arises from the microscopic atomic length scale physics is incompletely understood. This interaction will become more and more important for increasing doping concentration.
- Coulomb interactions between free carriers and Mn acceptors and among the Mn acceptors are also important. For dilute Mn systems, the free carriers will reside in acceptor-bound impurity bands. In ignoring these interactions in our simplest model we are appealing to the metallic nature of the free carrier system at larger Mn concentrations, that will result in screening. We believe that this source of disorder will be important in determining the $T = 0$ resistivity and the magnetic anisotropy energy.
- Including Coulomb interaction among the itinerant carrier enhances ferromagnetism as shown in mean-field calculations [33].
- Microscopic details of the p-d exchange physics suggest a finite effective range of the exchange interaction [24]. In fact, a finite range is also required as an ultraviolet cutoff since a delta-function type of interaction with discrete Mn spins leads to an ill-defined problem with infinite negative ground-state energy. In this article we have circumvent this problem by assuming a homogeneous Mn density.

To estimate the additional effect of a finite exchange interaction range we replace the delta-function with a Gaussian function

$$J_{\text{pd}}(r) \rightarrow \frac{1}{(2\pi)^{3/2}l^3} e^{-r^2/(2l^2)} \tag{28}$$

with cutoff parameter l. We can redo the analysis of the spin waves and find that, in the long-wavelength limit, the dispersion is modified from $\Omega_k = Dk^2 + \mathcal{O}(k^4)$ to

$$\Omega_k = (D + \Omega^{\text{MF}}l^2)k^2 + \mathcal{O}(k^4), \tag{29}$$

i.e., the spin stiffness acquires an additive constant.

10   Conclusion

In conclusion we have presented some aspects of a theory a ferromagnetism in doped diluted magnetic semiconductors which goes beyond both the RKKY picture and beyond the mean-field approximation for the kinetic-exchange model. The present discussion has focussed on the dependence of the system's collective spin excitations on model parameters and on implications for long-range magnetic order.

Acknowledgements

We thank M. Abolfath, D. Awschalom, B. Beschoten, A. Burkov, T. Dietl, J. Furdyna, S. Girvin, T. Jungwirth, B. Lee, H. Ohno, and J. Schliemann for useful discussions. This work was supported by the Deutsche Forschungsgemeinschaft under grant KO 1987-1/1, by the National Science Foundation, DMR-9714055, and by the Indiana 21st Century Fund.

References

1. G.A. Prinz, Physics Today **48**, April issue, 58 (1995).
2. G.A. Prinz, Science **282**, 1660 (1998).
3. J.M. Kikkawa, I.P. Smorchkova, N. Samarth, and D.D. Awschalom, Science **277**, 1284 (1997).
4. J.M. Kikkawa and D.D. Awschalom, Phys. Rev. Lett. **80** 80, 4313 (1998).
5. J.M. Kikkawa and D.D. Awschalom, Nature **397**, 139 (1999).
6. G. Schmidt, D. Ferrand, L.W. Molenkamp, A.T. Filip, and B. J. van Wees, Phys. Rev. B **62**, R4790 (2000).
7. R. Fiederling, M. Keim, G. Reuscher, W. Ossau, G. Schmidt, A. Waag, and L.W. Molenkamp, Nature **402**, 787 (1999).
8. Y. Ohno, D.K. Young, B. Beschoten, F. Matsukura, H. Ohno, and D.D. Awschalom, Nature **402**, 790 (1999).
9. J.K. Furdyna and J. Kossut, *Diluted Magnetic Semiconductors*, Vol. 25 of *Semiconductor and Semimetals* (Academic Press, New York, 1988).

10. T. Dietl, *Diluted Magnetic Semiconductors*, Vol. 3B of *Handbook of Semiconductors*, (North-Holland, New York, 1994).
11. A. Haury, A. Wasiela, A. Arnoult, J. Cibert, S. Tatarenko, T. Dietl, Y. Merle d' Aubigné, Phys. Rev. Lett. **79**, 511 (1997).
12. H. Ohno, H. Munekata, T. Penney, S. von Molnár, and L.L. Chang, Phys. Rev. Lett. **68**, 2664 (1992).
13. H. Ohno, A. Shen, F. Matsukura, A. Oiwa, A. Endo, S. Katsumoto, and Y. Iye, Appl. Phys. Lett. **69**, 363 (1996).
14. H. Ohno, Science **281**, 951 (1998).
15. H. Ohno, J. Magn. Magn. Mater. **200**, 110 (1999).
16. T. Hayashi, M. Tanaka, K. Seto, T. Nishinaga, H. Shimada, and K. Ando, J. Appl. Phys. **83**, 6551 (1998).
17. T.M. Pekarek, B.C. Crooker, I. Miotkowski, and A.K. Ramdas, J. Appl. Phys. **83**, 6557 (1998).
18. A. Van Esch, J. De Boeck, L. Van Bockstal, R. Bogaerts, F. Herlach, and G. Borghs, J. Phys. Condensed Matter **9**, L361 (1997).
19. A. Van Esch, L. Van Bockstal, J. De Boeck, G. Verbanck, A.S. van Steenbergen, P.J. Wellmann, B. Grietens, R. Bogaerts, F. Herlach, and G. Borghs, Phys. Rev. B **56**, 13103 (1997).
20. A. Oiwa, A. Endo, S. Katsumoto, Y. Iye, H. Ohno, and H. Munekata, Phys. Rev. B **59**, 5826 (1999).
21. F. Matsukura, H. Ohno, A. Shen, and Y. Sugawara, Phys. Rev. B **57**, R2037 (1998).
22. B. Beschoten, P.A. Crowell, I. Malajovich, D.D. Awschalom, F. Matsukura, A. Shen, and H. Ohno, Phys. Rev. Lett. **83**, 3073 (1999).
23. The central cell corrections required for an accurate envelope-function description of an isolated Mn^{2+} ion in GaAs [24] will be reduced by free-carrier screening.
24. A.K. Bhattacharjee and C. Benoit à la Guillaume, Solid State Commun. **113**, 17 (2000).
25. M. Sigrist, H.Tsunetsugu, and K. Ueda, Phys. Rev. Lett. **67**, 2211 (1991).
26. A. Babcenco and M.G. Cottam, Solid State Commun. **22**, 651 (1977).
27. S. Doniach and E.P. Wohlfarth, Proc. R. Soc. London Ser. A **296**, 442 (1967).
28. J. König, H.H. Lin, and A.H. MacDonald, Phys. Rev. Lett. **84**, 5628 (2000).
29. J. König, H.H. Lin, and A.H. MacDonald, to be published in the Proceedings of "The International Conference on the Physics and Application of Spin-Related Phenomena in Semiconductors", Sendai (Japan), September 2000, Physica E.
30. J. König, H.H. Lin, and A.H. MacDonald, to be published in the Proceedings of the "25th International Conference on the Physics of Semiconductors" Osaka (Japan), September 2000.
31. H. Ohno, N. Akiba, F. Matsukura, A. Shen, K. Ohtani, and Y. Ohno, Appl. Phys. Lett. **73**, 363 (1998).
32. T. Dietl, A. Haury, and Y.M. d'Aubigné, Phys. Rev. B **55**, R3347 (1997).
33. T. Jungwirth, W.A. Atkinson, B.H. Lee, and A.H. MacDonald, Phys. Rev. B **59**, 9818 (1999).
34. T. Dietl, H. Ohno, F. Matsukura, J. Cibert, and D. Ferrand, Science **287**, 1019 (2000).
35. B.H. Lee, T. Jungwirth, and A.H. MacDonald, Phys. Rev. B **61**, 15606 (2000).
36. M. Abolfath, J. Brum, T. Jungwirth, and A.H. MacDonald, cond-mat/0006093.
37. T. Dietl, H. Ohno, and F. Matsukura, cond-mat/0007190.
38. J. Schliemann, J. König, H.H. Lin, and A.H. MacDonald, cond-mat/0010036.
39. A. Auerbach, *Interacting Electrons and Quantum Magnetism* (Springer, New York, 1994).
40. J.M. Luttinger and W. Kohn, Phys. Rev. **97**, 869 (1955).

Part VI

Nanomechanics

Nano-Electromechanical Systems: Displacement Detection and the Mechanical Single Electron Shuttle

R.H. Blick, F.W. Beil, E. Höhberger, A. Erbe, and C. Weiss

Center for NanoScience and Sektion Physik, Ludwig-Maximilians-Universität, Geschwister-Scholl-Platz 1, 80539 München, Germany.

Abstract. For an introduction to nano-electromechanical systems we present measurements on nanomechanical resonators operating in the radio frequency range. We discuss in detail two different schemes of displacement detection for mechanical resonators, namely conventional reflection measurements of a probing signal and direct detection by capacitive coupling via a gate electrode. For capacitive detection we employ an on-chip preamplifier, which enables direct measurements of the resonator's displacement. We observe that the mechanical quality factor of the resonator depends on the detection technique applied, which is verified in model calculations and report on the detection of sub-harmonics. In the second part we extend our investigations to include transport of single electrons through an electron island on the tip of a nanomachined mechanical pendulum. The pendulum is operated by applying a modulating electromagnetic field in the range of 1 − 200 MHz, leading to mechanical oscillations between two laterally integrated source and drain contacts. Forming tunneling barriers the metallic tip shuttles single electrons from source to drain. The resulting tunneling current shows distinct features corresponding to the discrete mechanical eigenfrequencies of the pendulum. We report on measurements covering the temperature range from 300 K down to 4.2 K. The transport properties of the device are compared in detail to model calculations based on a Master-equation approach.

1   Introduction

Nano-electromechanical systems (NEMS) are investigated because of their promising features regarding sensitive tools for sensor and communications technology. This systems may also be used as 'quantum-mechanical' resonators which allow to explore mechanics in the quantum mechanical range when operated at several GHz at ultra low temperatures. In the first part we want to address different detection schemes necessary to achieve sensitive displacement detection for NEMS. It is shown that the measured Q-value of this structures depends on the detection scheme used. This fact permits to increase the sensitivity of those systems by using the appropriate detection technique. In this way a handle on quantum squeezing experiments with these mesoscopic mechanical systems is attained [1]. This scheme of detection is also suited for probing quadrature squeezed states [2].

In the second part we will focus on a non-classical version of one of the traditional experiments in the electrodynamics class: Usually it is set up by two large capacitor plates and a metallized ball suspended in between the plates.

Applying a constant voltage of some 100 V across the plates leads to the onset of periodic charge transfer by the ball bouncing back and forth, similar to a classical bell [3]. The number of electrons transferred by the metallized ball in each revolution naturally depends on the volume of the metal, but can be estimated to be of the order of 10^{10}. At an oscillation frequency of some 10 Hz up into the audible kHz-range this gives a typical current of $1-10$ μA. The question arising is whether such an experiment can be performed on the microscopic level in order to achieve a transfer not of a multitude but of only one electron per cycle of operation at frequencies of some 100 MHz. Indeed this can be achieved by simply scaling down the setup and applying a nanomechanical resonator. Such a device is of great importance for signal processing applications and for metrology, since it allows to transfer single electrons at radio frequencies and reduces cotunneling events at the same time.

2   Fabrication of Nanoscale Device with Three-Dimensional Relief

The beam-resonators are built up of freely suspended silicon beams, covered by an 50 nm resp. 100 nm thick conducting layer, close to which tapered electrodes are mounted sidewise. These electrodes may serve for capacitive excitation and detection purposes. During the fabrication process optic and electron lithographic steps are used first, to structure gates and an etch mask. Than the resonators are defined by succeeding steps of dry and wet etching. Electron micrographs of this structures are show in Fig. 6. Processing of the electron shuttle is more advanced, since two steps in electron beam lithography have to be performed. The mechanical pendulum itself is defined by etching in a combination of dry and wet etch steps. Alignment of the etch mask with respect to the metallic leads has to be accurate down to 10 nm in order to provide well defined tunneling contacts. In earlier work we ensured that the processing steps, which are also used in our present work, provide clean and tuneable tunneling contacts [4]. A schematic drawing of the measurement setup is shown in Fig.6. The details of the techniques used to fabricate the nanoscale devices are reported in preceding work in more detail [5–7].

3   Detecting Motion of Nanomechanical Resonators

The resonators can be driven either by the capacitive forces of the electrodes, or by Lorentz forces, due to placement of the structures in an external magnetic field and inducing an alternating current along the length of the conducting metal on top of the beam. One way of detecting the amplitude of an oscillating motion is to take advantage of the amplitude dependent impedance $\hat{Z}_{res}$ of the resonator component, due to electro motive forces. This is easily measured with an usual networkanalyser $HP8751A$ together with a test-set in order to measure the S-parameters of the component. With the mentioned side-gates it

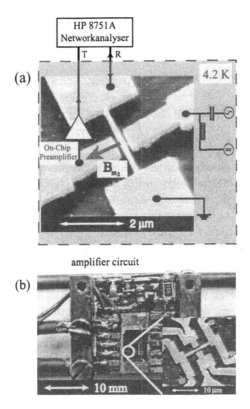

Fig. 1. (a) Scanning electron microscope micrograph of one of the nanomechanical resonators. The silicon beam is covered by a thin Au-sheet; the electrodes on the left and right allow application of an additional ac voltage modulation. Also pictured is the experimental setup for sampling the mechanical properties of the suspended beam. For characterization we employ a network analyzer scanning the frequency ω over the range of interest (see text for details). (b) On-chip amplifier: One of the gates is coupled to the amplifier enabling capacitive detection of the beam's displacement. A magnified view of one of these is depicted in the inset.

is also possible to measure the mechanical displacement of the oscillating part capacitive. Due to the very small signal currents we used for this measurements a on-chip preamplifier, which was incorporated with the above mentioned high-speed field effect transistor, which served as an impedance-converter. An aerial view of the setup and the amplifier is depicted in Fig. 1. The magnetic field orientation was chosen to be parallel to the surface of the transistor. The total impedance of the beam resonator placed in a magnetic field B is derived in [8]. This can be simplified assuming harmonic excitation at the eigenfrequency:

$$\hat{Z}_{res}(\omega = \omega_0) = R + \frac{L^2 B^2}{2\mu m_{eff}}, \tag{1}$$

where m_{eff} is the effective mass of the beam and the length of the beam l is connected via $L = l\pi/2$. The attenuation constant is referred to as μ. The minimum detectable force using mechanical cantilevers is limited by the vibrational noise and is given by [8]

$$F_{min} \approx \sqrt{\frac{\kappa}{\omega_0}} \approx \sqrt{\frac{wt^2}{lQ}}, \qquad (2)$$

where κ is the spring constant of the beam or cantilever, w, t, l, are the width, thickness, and length of the beam and Q is the mechanical quality factor, being defined as $Q = f_0/\delta f_0$, where δf_0 denotes the width of the resonance peak. Regarding this context the aim is to achieve a considerable size reduction of the structures leading to higher eigenfrequencies.

Fig. 2 shows measured spectra of the reflected power for one of the two different magnetically driven beams with different dimensions (henceforth these resonators will be called m_1 and m_2, as assigned in Fig. 2). This resonators are for $4.2K$ resonant at $95.93MHz$ for m_1 resp. $81.7MHz$ for m_2, where the varied magnetic field was oriented perpendicular to the beam for m_1 and parallel for m_2, so that two different modes, were generated. These resonances could be verified by capacitive detection. There are several factors which limit the quality of the resonator. The most important are losses due to the ohmic resistance of the conduction layers, the viscous inner damping of the material itself and the external resistance of the setup. The upper bound for the Q-value due to an finite resistance of the conduction layer can easily be estimated. Using this setup as an electrometer yields a charge sensitivity of 1.3×10^{-3} e/$\sqrt{Hz}$, which is three orders of magnitude better than previously measured [10].

The result when using capacitive detection to probe the resonances of m_2 is depicted in Fig. 3. In this measurement we observed an enhanced quality factor for m_2 of $Q \sim 4150$. This value is increased by a factor of 1.52 as compared to the value measured by the standard detection technique. This behaviour can be modeled by simply taking into account the different functional dependencies of the measuered signal on the excitation frequency [8]. The resonance has Lorenzian shape when measuring the induced voltage. Using capacitive detection the signal depends on the square of the Lorenzian, resulting in an enhanced quality-factor and a shift of the peaks. The factor of 1.52 for the enhanced Q-value is reproduced by the calculations while the shift is observed but too large compared with theory.

To further increase the sensitivity of NEMS for sensor applications a simple approach is to probe harmonics of the base frequency at higher frequencies. Although, it is easy to calculate resonance frequencies by finite element methods (FEM), there has been no observation of several modes for a single structure so far. In Fig. 3 we present measurements on a third beam with sligthly modified dimensions revealing two Lorenzians. The measured spectrum in Fig. 4 shows two resonances at 26.9 MHz and 35.5 MHz. Using FEM [11] we can assign the resonance at $\omega_0 = 35.6$ MHz to the first eigenmode of the beam, which is also

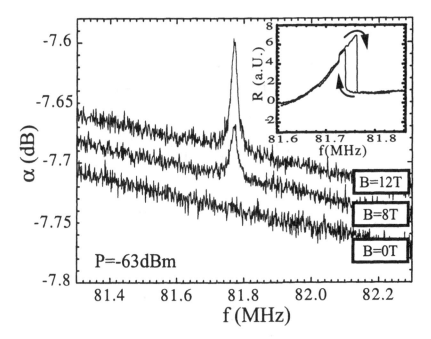

Fig. 2. Measured resonances for the two different beams. This curves where measured scanning the reflected power over the frequency-range of interest for different applied magnetic fields. This mechanical resonator is 4.8 μm long, 170 nm width and 190 nm thick, covered by an 50 nm thick conducting Au layer. The irradiated power amounted $-63dB$. The elongation of the structure is estimated to be about $10nm$. The inset shows nonlinear effects observed when scanning the frequnecy range from low to high frequencies and reverse. In this case a hysteresis emerges.

depicted in the inset of Fig. 4. The second resonance at 27 MHz can be explained by resonant excitation of a subharmonic at $\omega = 3/4 \times \omega_0$. The inset of the Fig. 4 also shows the Q-values of the resonances for several temperatures.

4   How to Shuttle Single Electrons Mechanically

In this second part we want to focus on our results of ours on shrinking a mechanical electron shuttle to submicron dimensions by integration of an electron island into a nanomechanical resonator functioning as an electromechanical transistor (EMT) [12]. The clear advantages are the increased speed of operation and the reduction of the transfer rate, allowing to count electrons one by one. This combination was proposed theoretically by Gorelik et al. [13] for metallic particles, which are connected to the reservoirs by elastically deformable organic molecu-

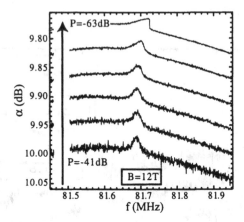

Fig. 3. Resonances in the spectrum of m_2 detected by capacitive coupling.

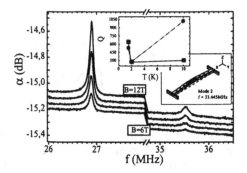

Fig. 4. Radio frequency spectrum of the reflected power for another resonator: Shown is the resonance of the base frequency $f_0 = 35.6$ MHz and a subharmonic at $3/4 \times f_0$. In the insets the calculated modeshape of the corresponding eigenmode is shown as well as the variation of the Q-value with temperature for the two peaks observed. The square dots refer to the resonance at 37 MHz while the circular dots refer to the 35.6 MHz peak.

lar links. The main difference to common SET devices is the fact that only one tunneling barrier is open at a certain time. This naturally leads to a suppression of cotunneling effects and thus increases the accuracy of current transport [14]. The reduced cotunneling effects may lead to accurate measurements of quantum fluctuations in these devices as well.

An important feature of the device is the possibility to effectively modulate the tunneling rate onto and off the electron island given by the large speed of operation $f \sim 100$ MHz. This basically enables to mechanically filter and select electrons passing through the electromechanical circuit by simply adjusting the tunneling rate Γ. Additionally, the device is a tool to regularize the stochastic tunneling process through a well defined geometry. Another advantage of nanomechanical resonators, machined out of Silicon On Insulator (SOI) material, is their insensitivity to thermal and electrical shocks as has been shown in their application for electrometry [15,16]. This and the high speed of operation enable direct integration in filter applications [17]. We have already demonstrated, that a nanomechanical tunneling contact, which operates at radio frequencies (rf), can be built out of SOI substrates [4]. The tunneling process in turn is very sensitive to changes of the environmental conditions, thus its use in sensor applications.

A schematic drawing of the measurement setup is shown in Fig. 6. In order to guarantee clean tunneling contacts for the real device shown in the sample chamber was heated and pumped prior to the measurements. The first measurements were performed at room temperature. The sample was mounted in an evacuated sample holder with a small amount of helium gas added, to ensure thermal coupling. The 300 K trace shows a variety of resonances, where the source/drain current is increased due to the motion of the clapper. This behavior is well known from the measurements performed on the single tunneling barrier [4].

In order to calculate the probability $p(m, t)$ to find m additional electrons on the island at time t a Master equation was used [14]. In our case the island oscillates mechanically ($x(t) = x_{\max} \sin(\omega t)$) which leads to time dependent transition rates $\Gamma(m, t)$ at the leads. Collecting gain and loss terms the Master equation reads

$$\frac{d}{dt} p(m, t) = - \left[\Gamma_{\mathrm{L}}^{(+)}(m, t) + \Gamma_{\mathrm{R}}^{(+)}(m, t) + \Gamma_{\mathrm{L}}^{(-)}(m, t) + \Gamma_{\mathrm{R}}^{(-)}(m, t) \right] p(m, t)$$
$$+ \left[\Gamma_{\mathrm{L}}^{(+)}(m - 1, t) + \Gamma_{\mathrm{R}}^{(+)}(m - 1, t) \right] p(m - 1, t)$$
$$+ \left[\Gamma_{\mathrm{L}}^{(-)}(m + 1, t) + \Gamma_{\mathrm{R}}^{(-)}(m + 1, t) \right] p(m + 1, t), \qquad (3)$$

with golden rule tunneling rates of the form [18]

$$\Gamma = \frac{1}{e^2 R} \frac{\Delta E}{1 - \exp\left(-\frac{\Delta E}{k_{\mathrm{B}} T}\right)}. \qquad (4)$$

The time dependence of both R and the capacitance C in $\Delta E \propto \frac{e^2}{C}$ leads to

$$\Gamma_{\mathrm{R}}^{(\mp)}(m, t) = g_{\mathrm{R}}(t)\, \Gamma_{\mathrm{R}, m}^{(\mp)}(t). \qquad (5)$$

Here

$$g_{\mathrm{R}}(t) = \frac{R_{\mathrm{R}}(t_{\max}) C_{\mathrm{R}}(t_{\max})}{R_{\mathrm{R}}(t) C_{\mathrm{R}}(t)} \qquad (6)$$

is a strongly varying function of time, which is dominated by the exponentially varying $R(t)$ ($t_{\max}$ is the time where the island is at its closest point to the right electrode). For the rates $\Gamma_{\mathrm{R}, m}^{(\mp)}(t)$ whose time dependence is determined by that of the capacitance only, we take the standard result for tunneling rates when a voltage of $-V/2$ is applied at the left electrode and a voltage of $V/2$ is applied at the right electrode:

$$\Gamma_{R,m}^{(\mp)}(t) = \frac{1}{\tau} \frac{\pm\left(m + \frac{C_L(t)V}{e}\right) - \frac{1}{2}}{1 - \exp\left[-\left(\pm\left(m + \frac{C_L(t)V}{e}\right) - \frac{1}{2}\right)\frac{e^2}{C_\Sigma(t)k_B T}\right]} \tag{7}$$

where $\tau = R_R(t_{max})C_R(t_{max})$ and $C_\Sigma(t) = C_R(t) + C_L(t)$. For the left electrode the indices R and L have to be interchanged, V has to be replaced by $-V$ and $x(t)$ by $-x(t)$.

In order to be able to solve the master equations analytically, we replace the function $g_R(t)$ by a step function $\tilde{g}_R(t)$ having the same height and area:

$$\tilde{g}_R(t) \equiv \begin{cases} 0 & : \quad t \le t_{max} - t_0 \\ 1 & : \quad t_{max} - t_0 < t < t_{max} + t_0 \\ 0 & : \quad t \ge t_0 + t_{max} \end{cases}, \qquad t_0 \equiv \sqrt{\frac{\pi\lambda}{2x_{max}}}\frac{1}{\omega}\left(1 + \frac{\lambda}{2x_0}\right). \tag{8}$$

The width $2t_0$ corresponds to an effective contact time. As expected, the contact time decreases for increasing oscillation frequencies ($x_{max} \gg \lambda$, $t_{max} \gg t_0$).

The solutions $p(m|n,t)$ of the Master equation can be used to calculate the probability $p_\Delta(m)$ that m electrons are transferred:

$$p_\Delta(n) \equiv \sum_k p_i(k)p(k - n|k, t_0 + t_{max}), \tag{9}$$

where $p_i(k)$ is the initial probability to find k additional electrons on the island ($\sum_k p_i(k) = 1$). From $p_\Delta(n)$ we then calculated both $\langle N \rangle$ and ΔN ($\langle N^k \rangle \equiv \sum_N N^k p_\Delta(N)$).

For very low temperatures $k_B T \ll \frac{e^2}{C}$ Coulomb blockade fixes the number of additional electrons n on the island between $-n_{max}$ and n_{max} strictly. The resulting coulomb staircase is shown in Fig. 5.

In the middle of the first step the analytic expressions for the average number of electrons transferred ($\langle N \rangle$) and the mean square fluctuations ($(\Delta N)^2$) are comparatively simple [14]:

$$\langle N \rangle \quad = 2\frac{1 - a^3}{[1 + a]\left[1 + \frac{1}{2}a + a^2\right]} \tag{10}$$

$$(\Delta N)^2 = 2a(1 - a)\frac{6 + 9a + 22a^2 + 13a^3 + 10a^4}{\left[2a^2 + a + 2\right]^2 [a + 1]^2} \tag{11}$$

where $a \equiv \exp\left(-\frac{t_0}{\tau}\right)$.

In the present experiment [12] the clapper is set into motion by an ac-voltage applied to the two driving gates at frequency f_0. This leads to an alternating force acting on the grounded lower part of the pendulum. Additionally, the driving voltage also acts as a gate voltage. The gate capacitance is approximately

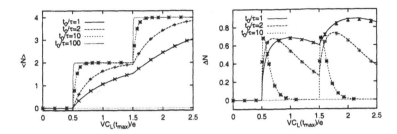

Fig. 5. The average number of electrons transferred per period (left) and the root mean fluctuations (right) for $T = 0$. Coulomb blockade is clearly visible: up to a critical voltage ($\frac{VC_L(t_{max})^{\cdot}}{e} = \frac{1}{2}$) no electrons are transferred. The Coulomb staircase becomes symmetric for $t_0 \gg \tau$. The agreement between analytical results (lines) and computer simulations (crosses) is excellent [14].

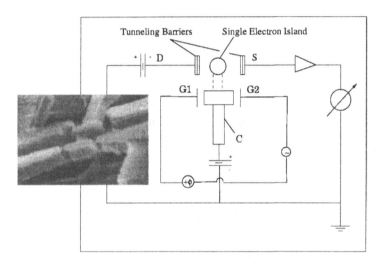

Fig. 6. Schematic drawing of the measurement setup. The single electron island is situated at the top of the clapper and supported by a silicon beam. The clapper is set into motion by applying an ac-voltage to the driving gates G_1 and G_2. The ac-voltage on the gates are phase shifted for optimal coupling of the ac-power to the motion of the beam. The dc-current from the source to the drain contact is measured with a current-voltage converter. The inset shoes an SEM-picture of the sample.

$C \approx 84$ aF which corresponds to gate charges of up to $\pm 527\,e$ if voltages of up to $\pm 1V$ are applied. We modified Eq. (3) in order to account for the influence of the gate voltage [12]. If the frequency f_0 of the driving force meets the eigenfrequency of the clapper, resonant motion is excited. In our calculations this kind of motion shows a large number of electrons transferred at a high accuracy during each cycle. This creates the large peaks seen in the measurements. If the shuttle moves with frequencies different from f_0 (*e.g.* excited thermally), the resulting current has a much smaller signal to noise ratio. Therefore, the large value of the gate-voltage explains both the high number of electrons transferred at room temperature and the background depending on whether or not the clapper moves with the same frequency as the applied voltage.

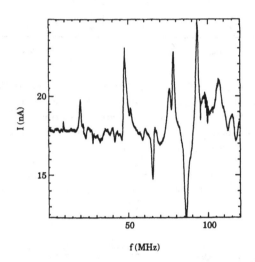

Fig. 7. Measurement of the tunnel current from source to drain at room temperature. The great number of peaks can be understood by assuming a complex mode spectrum. This fact is supported by numerical calculations using a finite element program [11].

Numerical solution of the modified Eq. (3) results in the height of the peaks being determined by the applied gate–voltage V_G. The number of electrons transferred does hardly depend on source/drain bias, since the voltages applied on the driving gates are a factor 10^3 larger than the source/drain bias. In the experiment the peaks are superimposed on a background, which depends linearly on the source/drain bias. This background is due to the thermal motion of the clapper, since it disappears at lower temperatures. The displacement noise of a cantilever can be calculated in analogy to an electrical circuit [19]. The response of the clapper to an external force was calculated by a finite element program [11]. We conclude from this calculation, that contributions to the current can be expected up to frequencies of some GHz. This explains the amplitude of the thermal background.

The peak height does not depend on the dc-source/drain bias as well. Both facts can be explained by the large gate voltage.

The number of transferred electrons in this peak is ≈ 1000 which agrees with our theoretical estimate based on Eq. (3). These results have shown, that the electron transfer works well at room-temperature. Since electron tunneling should be very sensitive to environmental influences this opens an interesting possibility for sensor applications.

Measurements at lower temperatures show a complete suppression of the ohmic background and thus indicate its thermal nature. At temperatures of about 12 K a pronounced peak at 120 MHz is found. This peak is strongly attenuated towards lower temperatures of 4.2 K due to the increased stiffness of the clapper. The rest of the complex spectrum is completely suppressed at 4.2 K also because of the increased stiffness of the structure.

At helium temperature the motion of the clapper is strongly reduced, resulting in very high tunnel barriers at the turning points. The maximum current amplitude of the peak is (2.3 ± 0.02) pA, which corresponds to a transfer of 0.11 electrons on average per cycle of motion of the clapper. Peaks in the classical experiment [4] show Lorentzian line shape. In order to obtain a formula for the peak fit for the present experiment we use the expression 10 in the limit of low contact times. Combined with the well known equations for the damped harmonic oscillator

$$x''(t) + 2\pi k x'(t) + 4\pi^2 f_0^2 x(t) = F \sin(2\pi f t) \tag{12}$$

we obtain [12]

$$N = \frac{A}{f} \sqrt{\frac{x_{max}(f_r)}{x_{max}(f)}} \, \exp\left[-B\left\{1 - x_{max}(f)/x_{max}(f_r)\right\}\right] \tag{13}$$

where $x_{max}(f) \propto 1/\sqrt{\left(f^2 - f_0^2\right)^2 + k^2 f^2}$ is the amplitude of the oscillation in resonance and $f_r = \frac{f_0}{2}\sqrt{4 - 2k}$ is the shifted frequency of the damped oscillator, respectively. From the fit parameters we obtain a quality factor $Q = \frac{f_0}{k}$ of order 10. Small Qs are essential for operation as a switch where the oscillating force in Eq. (12) is replaced by a step function. For small quality factors the oscillatory solution of the differential equation vanishes on a short timescale.

The shape of the resonance differs strongly from the measurements on a single tunnelbarrier shown in Fig.8 b). In these measurements the peak shape could be modelled by a Lorentzian line shape [4].

5   Summary

We have presented measurements on several nanomachined resonators operating in the radio frequency regime. An on-chip preamplifier enables us to detect the displacement of the nanowires directly by capacitive coupling. This is compared to the conventional method which monitors the reflection of incident power. We find changes in the Q-factors depending on the detection scheme applied. This dependence can be modelled when taking into account the difference in dependence of the signals on the amplitude of oscillation. We also found evidence for the generation of sub-harmonic resonances of the nanomachined beam studied. This will allow to further increase the force sensitivity, by pumping the nanomechanical system on the fundamental mode while probing on one of the harmonic modes.

a)   b)

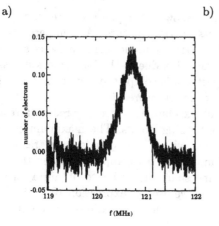

f (MHz)

Fig. 8. a) At low temperatures only one peak remains, which can be fitted by the model given in Eq. (13). b) For comparison we show a peak measured on a single mechanically moving tunneling contact [4].

In further experiments on mechanical resonators we have shown single electron tunneling by using a combination of nanomechanics and single electron devices. We have demonstrated a new way to transfer electrons one by one at radio frequencies. At 4.2 K we measured an average of 0.11 ± 0.001 electrons which shows that the resolution of current transport through the shuttle should also resolve Coulomb blockade effects. We estimate the temperature, at which Coulomb blockade should be observable to be 600 mK. Scaling down the island size will increase this temperature.

6 Acknowledgements

We like to thank J.P. Kotthaus and W. Zwerger for support and M.L. Roukes and H. Pothier for detailed discussions. We acknowledge financial support by the Deutsche Forschungsgemeinschaft under contracts DFG-Bl-487/1-1 and DFG-Bl-487/3-1.

References

1. X. Hu and F. Nori, Phys. Rev. B **53**, 2419 (1996).
2. D. Rugar and P. Grütter, Phys. Rev. Lett. **67**, 699 (1991).
3. P. Benjamin, *The Intellectual Rise in Electricity* (Appleton, New York, 1895), p. 507.
4. A. Erbe, R.H. Blick, A. Tilke, A. Kriele, and J.P. Kotthaus, Appl. Phys. Lett. **73**, 3751 (1998).
5. H. Krömmer, A. Erbe, A. Tilke, S.M. Manus, and R.H. Blick, Europhys. Lett. **50**, 101 (2000).

6. L. Pescini, A. Tilke, R.H. Blick, H. Lorenz, J.P. Kotthaus, W. Eberhardt, and D. Kern, Nanotechnology **10**, 418 (1999).

7. A. Kraus, A. Erbe, R.H. Blick, Nanotechnology **11**, 165 (2000); A. Kraus, Diploma thesis, Ludwig-Maximilians-Universität, Munich, Germany 2000.

8. F.W. Beil, A. Kraus, A. Erbe, E. Höhberger, and R.H. Blick, "Approching the limits of displacement in nanomechanical resonators", preprint (2000).

9. T.D. Stowe, K.Yasumura, T.W. Kenny, D.Botkin, K. Wago, D. Rugar, Appl. Phys. Lett. **71**, 288 (1997).

10. A.N. Cleland and M.L. Roukes, Nature **392**, 160 (1998).

11. solvia, v. 95.2 (a finite element system).

12. A. Erbe, C. Weiss, W. Zwerger, and R.H. Blick, "A nanomechanical resonator shuttling single electrons at radio frequencies", preprint (2000).

13. L.Y. Gorelik, A. Isacsson, M.V. Voinova, R.I. Shekter, and M. Jonson, Phys. Rev. Lett. **80**, 4526 (1998)

14. C. Weiss and W. Zwerger, Europhys. Lett, **47**, 97 (1999).

15. A.N. Cleland and M.L. Roukes, Appl. Phys. Lett. **69**, 2653 (1996).

16. H. Kroemmer, A. Erbe, A. Tilke, S.M. Manus, and R.H. Blick, Europhys. Lett. **50**, 101 (2000).

17. N.M. Nguyen and R. G. Meyer, IEEE J. of Solid-State Circuits, **SC-25**, no. 4, 1028 (1990).

18. *Single Charge Tunneling* edited by H. Grabert and M. H. Devoret, *NATO ASI Ser. B.* **294**, Plenum, New York (1992).

19. G. G. Yaralioglu and A. Atalar, Rev. Sci. Instr. **70**, 2379 (1999).

Lecture Notes in Physics

For information about Vols. 1–537
please contact your bookseller or Springer-Verlag

Monographs

For information about Vols. 1–24
please contact your bookseller or Springer-Verlag